# MISSILE GUIDANCE AI
## Kinematics, Dynamics and Control

There are three things which are too wonderful for me,
The way of an eagle in the air,
The way of a serpent on a rock,
The way of a ship in the midst of the sea.
*Book of Proverbs*


Armed with bows, using both the right hand and the left in hurling stones
and shooting arrows with the bow.
*Book of Chronicles*

# ABOUT THE AUTHOR

**Neryahu A. Shneydor** was born in Jerusalem, then Palestine, now Israel, in 1932. His secondary education was interrupted by the 1948 Independence War and the siege of Jerusalem. He was then 16 years old and became a junior member of the Haganah (the semi-legal military organisation) and joined the minuscule Signal Group of the city. He manned a signal station which used Morse-key operated lamps and, in daytime, heliographs, most of the equipment being of World War I vintage. This may have aroused his interest in electrical engineering. He studied Electrical Engineering in 1950-54 at the Technion – Israel Institute of Technology, Haifa, Israel, where he obtained his B.Sc. and Dipl. Ingénieur degrees. The latter was a relic from the time when Technion was under German influence: most of the founding professors being immigrants from Germany, some escapees from Nazism.

There followed four years of military services, mostly with the navy where he specialised in servo and in gunnery- and torpedo-fire control. Interest in these topics led to studies in control theory and the Master's degree. Dr. Shneydor joined RAFAEL, Israel Armament Development Authority, in 1960 - formerly a secret military unit called HEMED, Hebrew acronym for Science Corps. Among the early achievements of HEMED-RAFAEL were a radio-guided boat which became operational in 1955, and a sea-to-sea guided missile which evolved into the well-known Gabriel. Dr. Shneydor joined the servo group of RAFAEL, which specialised in mechanisms for controlling antennas, gimbals, actuators and other devices. He later participated in the development of Shafrir and Python air-to-air missiles. For his contribution to the development of the former's guidance system, he was awarded the Security Prize, Israel's highest token of appreciation.

In 1972-75 RAFAEL enabled him to study for a Doctoral degree, again at the Haifa Technion, under the supervision of the late Professor George Zames, with a thesis in nonlinear feedback control theory. In 1970 Dr. Shneydor had started to lecture on guidance and navigation at the Technion Department of Aeronautical Engineering (now the Faculty of Aerospace Engineering) and later on nonlinear control at the Department of Electrical Engineering as Adjunct Assistant Professor, being promoted to Adjunct Associate Professorship in 1979.

After 1975, Dr. Shneydor became occupied with research and development projects at RAFAEL, and in 1982-87 and 1992-94 he was R & D Deputy Director of the Missile Division. During his professional activity he published many reports, papers and texts on guidance. Now retired from RAFAEL, he continues lecturing on guidance and control and is a consulting engineer in these areas and studies the history of weapon guidance. Several papers on this topic have been presented at the Israel Annual Conferences on Aerospace Sciences.

# MISSILE GUIDANCE AND PURSUIT
## Kinematics, Dynamics and Control

**N.A. Shneydor**, B.Sc., Dipl. Ingénieur, M.Sc., Ph.D.
Israel Armament Development Authority
*and*
Senior Teaching Fellow
Technion – Israel Institute of Technology
Haifa, Israel

WOODHEAD
PUBLISHING

Oxford   Cambridge   Philadelphia   New Delhi

Published by Woodhead Publishing Limited,
80 High Street, Sawston, Cambridge CB22 3 HJ, UK
www.woodheadpublishing.com

Woodhead Publishing, 1518 Walnut Street, Suite 1100, Philadelphia,
PA 19102-3406, USA

Woodhead Publishing India Private Limited, G-2, Vardaan House, 7/28 Ansari Road,
Daryaganj, New Delhi – 110002, India
www.woodheadpublishingindia.com

First published by Horwood Publishing Limited, 1998; reprinted 2008
Reprinted by Woodhead Publishing Limited, 2011

British Library Cataloguing in Publication Data
A catalogue record for this book is available from the British Library

ISBN 978-1-904275-37-4

# Preface

Navigation has been with the human race from time immemorial. It is not surprising, therefore, that a very great number of books have been published on this ancient art. Guidance, on the other hand, has been first implemented, by building a remotely-guided unmanned boat for military purposes, in the beginning of this century. The technical literature on it is immense - articles, conference papers, reviews, bibliographies. However, surprisingly few books have been published that deal with guidance. If we do not count texts that are mainly descriptive, most of which appeared in the first decennary after World War II, we have less than half a dozen books in English. During my professional career as a research-and-development (R&D) engineer I also taught at a technical university and lectured for various industry and military audiences. I naturally used the existing texts, but gradually developed an approach which is different from theirs. Encouraged by colleagues and students, I eventually turned my lecture notes and transparencies into this book.

I believe this text differs from other ones in the field in several respects. Here are some of its key features.

⋆ Although it necessarilly emphasizes military applications of guidance, i.e., guided weapons, it also pays attention to guidance in nature: some real, some anecdotic, some invented by recreational mathematicians.

⋆ This book does not purport to be a history. However, it does try to give credit to pioneering scientists and to early developments and inventions.

⋆ In the theory of guidance one often has to solve differential and other equations. Wherever practical I present an analytic solution rather than resort to numerical ones: very often, the analytic solution enables one to discover interesting properties which would otherwise be obscured by lines of code and numerical data. Furthermore, most of the engineers and scientists among the readers, especially the younger ones, could easily make their own computer code wherever they wish to deepen their quantitative understanding of a problem studied.

⋆ Geometrical rules and guidance laws are stated in three-dimensional vector terms as well as the usual planar ones, and several examples have to do with three-dimensional guidance situations.

⋆ Many graphical illustrations are given of trajectories, launch zones and intercept zones, as well as of time histories of maneuver acceleration and other important variables. This should be of practical value for many readers.

This text is intended for people — students, engineers, analysts, physicists, programmers — involved or interested in any of the various aspects of guidance systems: use, development, design, manufacture, marketing, analysis, operational research. Mathematics at a first-year university level is the only prerequisite. However, for comprehending some portions of the text, acquaintance with feedback control theory would be helpful.

## Acknowledgements

I would like to thank the director of the Ecole des Mines de Paris, France, Dr. Jacques B. Levy, who made it possible for me to spend a sabbatical at the institute, when most of the work on this text was done, and Prof. Jean Lévine of the Centre Automatique et Systèmes of the Ecole des Mines for his invaluable aid during various stages of the work. I also wish to thank Prof. Aviv Rosen, Dean of the Faculty of Aerospace Engineering, Technion - Israel Institute of Technology, where I spent the last semester of the sabbatical. I was very fortunate to have the manuscript reviewed by several colleagues. The comments and suggestions made by Oded Golan and Ilan Rusnak of RAFAEL are gratefully acknowledged. I am especially indebted to Uri Reychav, also of RAFAEL, who devoted many long hours of his time during his sabbatical in the United States to going over the manuscript while it was being written in France. Particular thanks are extended to Rachel Weissbrod and Sarah Segev of RAFAEL library at Leshem and to Dr. Guy Shaviv of the Technion for their patient aid. Finally, I express my gratitude and love for my wife Na'ama, without whose support and encouragement this undertaking would not be possible.

N. A. Shneydor

Haifa, Israel
February 4, 1998

# Contents

Preface                                                                    v

Introduction                                                             xiii

1  Terminology and Definitions                                             1
   1.1  The Three Levels of the Guidance Process . . . . . . . . . . . .    1
        1.1.1  Definitions  . . . . . . . . . . . . . . . . . . . . . . .   1
        1.1.2  An Example  . . . . . . . . . . . . . . . . . . . . . .      3
        1.1.3  Scope of this Book . . . . . . . . . . . . . . . . . . .     3
   1.2  Terminology Related to Implementation . . . . . . . . . . . . .     3
        1.2.1  Definitions  . . . . . . . . . . . . . . . . . . . . . . .   3
        1.2.2  Examples . . . . . . . . . . . . . . . . . . . . . . . .     4
   1.3  Geometry and Kinematics . . . . . . . . . . . . . . . . . . . .     6
        1.3.1  Basic Definitions and Notations . . . . . . . . . . . . .    6
        1.3.2  Kinematics of Planar Motion  . . . . . . . . . . . . . .     7
   1.4  References . . . . . . . . . . . . . . . . . . . . . . . . . . .    8

2  Line-of-Sight Guidance                                                 11
   2.1  Background and Definitions . . . . . . . . . . . . . . . . . . .   11
   2.2  A Little History . . . . . . . . . . . . . . . . . . . . . . . .   12
   2.3  Kinematics  . . . . . . . . . . . . . . . . . . . . . . . . . .    14
        2.3.1  Analysis of the Planar Case . . . . . . . . . . . . . . .   14
        2.3.2  Examples for the Planar Case . . . . . . . . . . . . . .    16
        2.3.3  Remarks Concerning Practical Applications . . . . . . . .   23
        2.3.4  Kinematics in 3-D Vector Terms  . . . . . . . . . . . . .   25
        2.3.5  Kinematics of Modified LOS Guidance . . . . . . . . . . .   29
   2.4  Guidance Laws . . . . . . . . . . . . . . . . . . . . . . . . .    31
        2.4.1  Time-Domain Approach . . . . . . . . . . . . . . . . . .    33
        2.4.2  Classical-Control Approach . . . . . . . . . . . . . . .    36
        2.4.3  Optimal-Control Approach  . . . . . . . . . . . . . . . .   38

   2.5   Mechanization of LOS Guidance . . . . . . . . . . . . . . . . . . 39
        2.5.1   CLOS vs. Beam-Riding Guidance . . . . . . . . . . . . . 39
        2.5.2   On Tracking and Seekers . . . . . . . . . . . . . . . . . 41
        2.5.3   Mechanization in Practice . . . . . . . . . . . . . . . . . . 43
   2.6   References . . . . . . . . . . . . . . . . . . . . . . . . . . . . . . 44

**3  Pure Pursuit                                                47**
   3.1   Background and Definitions . . . . . . . . . . . . . . . . . . . 47
   3.2   Some of the Long History . . . . . . . . . . . . . . . . . . . . 48
   3.3   Kinematics . . . . . . . . . . . . . . . . . . . . . . . . . . . . . 49
        3.3.1   The Planar Case, Nonmaneuvering T . . . . . . . . . . 49
        3.3.2   The Planar Case, Maneuvering T . . . . . . . . . . . . . 57
        3.3.3   Other Interesting Planar Pursuits . . . . . . . . . . . . 58
        3.3.4   Deviated Pure Pursuit . . . . . . . . . . . . . . . . . . . 59
        3.3.5   Examples . . . . . . . . . . . . . . . . . . . . . . . . . . 67
   3.4   Guidance Laws for Pure Pursuit . . . . . . . . . . . . . . . . 69
        3.4.1   Velocity Pursuit vs. Attitude Pursuit . . . . . . . . . . 70
        3.4.2   A Simple Velocity-Pursuit Guidance Law . . . . . . . . 71
        3.4.3   A Simple Attitude-Pursuit Guidance Law . . . . . . . 73
   3.5   On the Mechanization of Pursuit Guidance . . . . . . . . . . 74
   3.6   References . . . . . . . . . . . . . . . . . . . . . . . . . . . . . 74

**4  Parallel Navigation                                       77**
   4.1   Background and Definitions . . . . . . . . . . . . . . . . . . . 77
   4.2   Kinematics of Planar Engagements . . . . . . . . . . . . . . 78
        4.2.1   Nonmaneuvering Target . . . . . . . . . . . . . . . . . . 78
        4.2.2   Maneuvering Target . . . . . . . . . . . . . . . . . . . . 82
        4.2.3   Variable Speed . . . . . . . . . . . . . . . . . . . . . . . 86
   4.3   Nonplanar Engagements . . . . . . . . . . . . . . . . . . . . . 87
        4.3.1   Definitions . . . . . . . . . . . . . . . . . . . . . . . . . 87
        4.3.2   Three properties . . . . . . . . . . . . . . . . . . . . . . 87
        4.3.3   Examples . . . . . . . . . . . . . . . . . . . . . . . . . . 88
   4.4   Guidance Laws for Parallel Navigation . . . . . . . . . . . . 92
        4.4.1   Proportional Navigation . . . . . . . . . . . . . . . . . . 92
        4.4.2   A Non-Feedback Law . . . . . . . . . . . . . . . . . . . 93
   4.5   Rules Related to Parallel Navigation . . . . . . . . . . . . . 95
        4.5.1   Constant Aspect Navigation . . . . . . . . . . . . . . . 95
        4.5.2   Constant Projected Line . . . . . . . . . . . . . . . . . . 96
   4.6   References . . . . . . . . . . . . . . . . . . . . . . . . . . . . . 99

**5  Proportional Navigation**                                               **101**
  5.1  Background and Definitions . . . . . . . . . . . . . . . . . . 101
  5.2  A Little History . . . . . . . . . . . . . . . . . . . . . . . . 103
  5.3  Kinematics of A Few Special Cases . . . . . . . . . . . . . . 104
      5.3.1  Two Special Values of $N$ . . . . . . . . . . . . . . . 104
      5.3.2  Stationary Target, Any $N$ . . . . . . . . . . . . . . . 104
      5.3.3  $N = 2$, Nonstationary, Nonmaneuvering Target . . . . . . . 105
  5.4  Kinematics of PN, Approximative Approach . . . . . . . . . . . 106
      5.4.1  True PN (TPN) . . . . . . . . . . . . . . . . . . . . 107
      5.4.2  Use of Range-Rate in TPN . . . . . . . . . . . . . . 109
      5.4.3  Pure PN (PPN) . . . . . . . . . . . . . . . . . . . . 109
      5.4.4  Some Results . . . . . . . . . . . . . . . . . . . . . 112
  5.5  Kinematics of PN, Exact Approach . . . . . . . . . . . . . . . 113
      5.5.1  TPN . . . . . . . . . . . . . . . . . . . . . . . . . 114
      5.5.2  PPN . . . . . . . . . . . . . . . . . . . . . . . . . 116
      5.5.3  TPN vs. PPN . . . . . . . . . . . . . . . . . . . . 117
  5.6  PPN and TPN in 3-D Vector Terms . . . . . . . . . . . . . . 118
      5.6.1  Definitions and Some Properties . . . . . . . . . . . . . 118
      5.6.2  An Example . . . . . . . . . . . . . . . . . . . . . 119
  5.7  Other Laws that Implement Parallel Navigation . . . . . . . . . 121
      5.7.1  Ideal PN . . . . . . . . . . . . . . . . . . . . . . . 122
      5.7.2  Prediction Guidance Law . . . . . . . . . . . . . . . 123
      5.7.3  Schoen's Laws . . . . . . . . . . . . . . . . . . . . 123
  5.8  References . . . . . . . . . . . . . . . . . . . . . . . . . . 124

**6  Mechanization of Proportional Navigation**                              **129**
  6.1  Background . . . . . . . . . . . . . . . . . . . . . . . . . . 129
  6.2  On the Structure of PN Systems . . . . . . . . . . . . . . . 129
  6.3  The Effects of Dynamics . . . . . . . . . . . . . . . . . . . 131
      6.3.1  Single-Lag Dynamics . . . . . . . . . . . . . . . . . 132
      6.3.2  Two-Lag Dynamics . . . . . . . . . . . . . . . . . . 134
      6.3.3  Higher-Order Dynamics . . . . . . . . . . . . . . . . 135
      6.3.4  The Stability Problem . . . . . . . . . . . . . . . . . 135
      6.3.5  Conclusions . . . . . . . . . . . . . . . . . . . . . 140
  6.4  Effects of Nonlinearities in the Guidance Loop . . . . . . . . . 140
      6.4.1  Variable Missile Speed . . . . . . . . . . . . . . . . 141
      6.4.2  Saturation of Lateral Acceleration . . . . . . . . . . . 144
      6.4.3  Saturations at the Seeker . . . . . . . . . . . . . . . 145
      6.4.4  Radome Refraction Error . . . . . . . . . . . . . . . 149
      6.4.5  Imperfect Stabilization of the Seeker . . . . . . . . . . 155
  6.5  Noise . . . . . . . . . . . . . . . . . . . . . . . . . . . . . 156
      6.5.1  Angular Noise . . . . . . . . . . . . . . . . . . . . 156

        6.5.2   Glint Noise . . . . . . . . . . . . . . . . . . . . . . 158
        6.5.3   Target Maneuver . . . . . . . . . . . . . . . . . . . 158
        6.5.4   Conclusions . . . . . . . . . . . . . . . . . . . . . . 159
        6.5.5   Remark . . . . . . . . . . . . . . . . . . . . . . . . 160
  6.6   References . . . . . . . . . . . . . . . . . . . . . . . . . . 160

**7  Guidance Laws Related to Prop. Navigation                    165**
  7.1   Background . . . . . . . . . . . . . . . . . . . . . . . . . 165
  7.2   PN Modified by Bias . . . . . . . . . . . . . . . . . . . . 166
        7.2.1   Augmented PN (APN) . . . . . . . . . . . . . . . . 166
        7.2.2   The Guidance-to-Collision Law . . . . . . . . . . . 168
  7.3   Guidance Laws for Low LOS Rates . . . . . . . . . . . . . 170
        7.3.1   Biased PN (BPN) . . . . . . . . . . . . . . . . . . 170
        7.3.2   Dead-Space PN . . . . . . . . . . . . . . . . . . . 171
  7.4   Proportional Lead Guidance (PLG) . . . . . . . . . . . . . 172
  7.5   Guided Weapons with Strapdown Seekers . . . . . . . . . 172
        7.5.1   An Integral Form of PN . . . . . . . . . . . . . . . 173
        7.5.2   Dynamic Lead Guidance (DLG) . . . . . . . . . . . 174
  7.6   Mixed Guidance Laws . . . . . . . . . . . . . . . . . . . . 175
        7.6.1   Mixed Guidance: PP and Parallel Navigation (or PN) . . . . 175
        7.6.2   Mixed Guidance: LOS Guidance and Other Laws . . . . . . 176
        7.6.3   Combining Midcourse Guidance and PN . . . . . . . . . 177
  7.7   References . . . . . . . . . . . . . . . . . . . . . . . . . . 178

**8  Modern Guidance Laws                                          181**
  8.1   Background . . . . . . . . . . . . . . . . . . . . . . . . . 181
  8.2   Methodology . . . . . . . . . . . . . . . . . . . . . . . . 182
  8.3   Principles of OCG, and Basic Examples . . . . . . . . . . 183
        8.3.1   Guidance and Optimal Control . . . . . . . . . . . 183
        8.3.2   OCG Laws for a Maneuvering Target . . . . . . . . . 185
        8.3.3   Laws for Systems with 1st Order Dynamics . . . . . . 187
        8.3.4   Laws for Systems with 2nd Order Dynamics . . . . . . 189
        8.3.5   Laws for Systems with High-Order Dynamics . . . . . 191
        8.3.6   A Short Summary . . . . . . . . . . . . . . . . . . 192
  8.4   A More General Approach to OCG . . . . . . . . . . . . . 194
        8.4.1   Definitions, and Statement of the Problem . . . . . . 194
        8.4.2   The LQ Problem . . . . . . . . . . . . . . . . . . . 195
        8.4.3   On the Solution to the LQ Problem . . . . . . . . . 196
        8.4.4   Two Examples . . . . . . . . . . . . . . . . . . . . 197
  8.5   Laws Based on LQG Theory . . . . . . . . . . . . . . . . 200
        8.5.1   Background . . . . . . . . . . . . . . . . . . . . . 200
        8.5.2   The LQG Problem . . . . . . . . . . . . . . . . . . 201

8.6 On the Mechanization of OCG Laws . . . . . . . . . . . . . . . 202
    8.6.1 Control Acceleration . . . . . . . . . . . . . . . . . . . 203
    8.6.2 Control Dynamics . . . . . . . . . . . . . . . . . . . . 204
    8.6.3 Radome Refraction Error . . . . . . . . . . . . . . . . 205
    8.6.4 Estimating the Time-to-Go . . . . . . . . . . . . . . . 205
    8.6.5 Estimating the System State . . . . . . . . . . . . . . 206
8.7 Comparison with Other Guidance Laws . . . . . . . . . . . . . 208
    8.7.1 OCG and Proportional Navigation . . . . . . . . . . . 208
    8.7.2 OCG and Other Modern Laws . . . . . . . . . . . . . 209
8.8 References . . . . . . . . . . . . . . . . . . . . . . . . . . . . . 211

**A  Equations of Motion                                217**
A.1 General . . . . . . . . . . . . . . . . . . . . . . . . . . . . . . 217
A.2 A Rotating FOC . . . . . . . . . . . . . . . . . . . . . . . . . 219
A.3 Coplanar Vectors . . . . . . . . . . . . . . . . . . . . . . . . . 220
A.4 Examples . . . . . . . . . . . . . . . . . . . . . . . . . . . . . 221

**B  Angular Transformations                           225**

**C  A Few Concepts from Aerodynamics                  229**
C.1 Skid-to-turn (STT) Configuration . . . . . . . . . . . . . . . . 229
C.2 Bank-to-turn (BTT) Configuration . . . . . . . . . . . . . . . 231
C.3 On Angle of Attack and Sideslip . . . . . . . . . . . . . . . . 231
C.4 Note . . . . . . . . . . . . . . . . . . . . . . . . . . . . . . . . 233

**D  Derivations of Several Equations                  235**
D.1 The Graphs of the $Kh$ Plane, Sec. 2.3.2 . . . . . . . . . . . . 235
D.2 Derivation of (2.21) . . . . . . . . . . . . . . . . . . . . . . . . 237
D.3 Proofs for (3.8) and (3.9) . . . . . . . . . . . . . . . . . . . . 237
D.4 On the $t_f$-Isochrones of Sec. 3.3.1(c) . . . . . . . . . . . . . . 238
D.5 Definition of DPP (Sec. 3.3.4) in Vector Terms . . . . . . . . . 239
D.6 A Proof for (4.11) . . . . . . . . . . . . . . . . . . . . . . . . 240
D.7 A Proof for Inequality (4.13) . . . . . . . . . . . . . . . . . . 240
D.8 Derivation of (4.15) of Sec. 4.2.2(a) . . . . . . . . . . . . . . . 241
D.9 Derivation Of (4.34) and (4.35) . . . . . . . . . . . . . . . . . 242
D.10 Vector Representation for Sec. 5.4.1 . . . . . . . . . . . . . . . 243
D.11 On Equivalent Noise Bandwidth . . . . . . . . . . . . . . . . . 244
D.12 APN Law in Vector Terms . . . . . . . . . . . . . . . . . . . . 244
D.13 Derivation of (8.14) . . . . . . . . . . . . . . . . . . . . . . . . 245
D.14 References . . . . . . . . . . . . . . . . . . . . . . . . . . . . . 245

**List of Symbols and Abbreviations                    247**
**Index                                                251**

# Introduction

According to the dictionary, 'guidance' is "the process for guiding the path of an object towards a given point, which in general may be moving." If the given point, which we will call the *target*, is fixed, e.g., a sea port, or its path in the future is known with sufficient accuracy, e.g., planet Mars, then the process is usually called *navigation*. If the target moves in a way that is not quite predictable — for example, a prey escaping its predator, an aircraft evading ground-to-air missiles — then the process is *guidance* in its narrower sense, which is the sense we will give it in this text.

The guided object may be a vehicle (a car, a boat, a missile, a spacecraft), a robot or, in fact, a living being. The process of guidance is based on the position and the velocity of the target relative to the guided object. The participants in the guidance process are also referred to in the literature as the *evader* and the *pursuer*, respectively. In nature, the ways predators catch their prey and some insects rendezvous their mates are certainly guidance processes. In human history, it is said that seamen, especially those who excercised the ignoble art of sea piracy, practised the rule we now call 'parallel navigation' (the 'navigation' part of the term being of course a misnomer) or 'collision course'. Mariners in general have known the inverse rule, which they apply in order to avoid collision at sea.

Modern, i.e., analytic, approach to guidance problems dates from the eighteenth century, when several mathematicians studied what we now call 'pure pursuit' or 'hound and hare pursuit'. This pursuit follows a very straightforward *geometrical rule*: run (or fly, or sail, as the case may be) where you see your target. Both this simple rule and the aforementioned parallel-navigation are *two-point guidance* rules, called so because only the pursuer and the target are involved in their respective definitions.

A family of geometrical rules for *three-point guidance* exists as well; the name derives from the fact that a third, reference point is required for the statement of the rule. In the most basic three-point geometrical rule, the pursuer is required to be on the line between the reference point and the target. For obvious reasons, this type of guidance is called 'line-of-sight guidance'.

Most of the applications for the theory of guidance are in weaponry. History for this kind of application begins in 1870, when Werner von Siemens submitted a proposal for "the destruction of enemy vessels by guided torpedos" to the Prussian ministry of war. Although not specifically said so by Siemens, the guidance of his proposed torpedo would have been of the line-of-sight type. We shall describe this proposal briefly later on; suffice it to mention now that by 1916 it had materialized into the first operational guided-weapon system in history.

The pure-pursuit rule was first applied to weapon systems in the early 1940's, during the second world war, when most of the basic relevant theory had in fact been known for two centuries and technical means for detecting targets and for conrolling guided vehicles had been developed. Towards the end of the war, a more sophisticated type of two-point guidance, called 'proportional navigation' for historical reasons, was studied. The basic theory of proportional navigation (PN) was first formulated in the United States in 1943. Some steps towards implementing a variant of PN in missile systems were taken in 1944 or 1945 by German scientists, who presumably did not know that the theory had already been developed elsewhere. The vast majority of two-point guided weapon systems existing today utilize PN in one of its numerous variants. There are nonmilitary applications of PN, too; for example, in space travel, extraterrestrial landing, and robotics.

PN has its limitations, though. In particular one should mention sensitivity to noise and to maneuvers carried out by the evader when the pursuer is approaching it. (To 'maneuver' means here to make abrupt changes in the direction of motion, i.e., execute high-acceleration turns; in pilots' parlance, to 'jink', and in mariners' one, to 'zigzag'.) A family of so-called 'modern guidance laws' has been developing since the early 1960's that do not suffer from these limitations or suffer much less. These laws are based on several recently developed techniques, in particular optimal-control theory and optimal-estimation theory, hence the often used terms 'optimal-control guidance' or just 'optimal guidance'.

This family of laws can be regarded as the most recent stage of the evolution process that started with Siemens's proposal. It seems that in spite of the maturity of the theory and the availability of the necessary technology, mostly microelectronics and computer science, practical application is still somewhat rare, probably due to economical reasons. Needless to say, however, the secrecy that prevails over armament development issues makes up-to-date, reliable information inaccessible, and therefore statements on recent developments are uncertain.

This is about as far as we go in this book. The next evolutionary stage would probably consist of laws based on differential-game theory. Although papers regarding this approach to guidance have been appearing since the 1970's, it seems that it is not ripe enough for inclusion in an introductory text like the present one.

The very fast progress of guided weaponry in the past fifty years would not be possible without advances in many technologies. One should mention internal-combustion engines, rocket motors, inertial instrumentation (especially gyroscopes), aeronautics, electronics (especially microelectronics and radar), electro-optics, and computer engineering. These and some other technological disciplines relevant to guidance are beyond the scope of this text, except where they have direct implications regarding its main topics. There are two reasons for this exclusuion. Firstly, including even some of the relevant disciplines would have made the book much weightier than what the author had in mind; secondly, an abundant literature is available that deals with most of the said technologies.

This book regards guidance from the point of view of the pursuer, i.e., how to arrive at the target, or intercept it. The inverse problem, that of avoidance, is not dealt with. Guidance is treated from the viewpoints of kinematics, dynamics, and control. In other words, we study trajectories, zones of interception, required maneuver effort, launch envelopes, stability of the guidance process, and related topics. Furthermore, technical problems involved with implementation and mechanization are discussed when they may affect accuracy, energy expenditure, and structural limits, hence, finally, costs.

The book is organized as follows. Following Chapter 1, which presents basic definitions and terminology, Chapters 2-7 deal with what have come to be called the *classical guidance laws*, namely
⋆ Line-of-sight guidance (Chapter 2),
⋆ Pure pursuit (Chapter 3),
⋆ Parallel navigation (Chapter 4),
⋆ Proportional navigation (Chapters 5 and 6),
⋆ Several guidance laws related to proportional navigation (Chapter 7).
Chapter 8 is dedicated to optimal-control guidance.

## REFERENCES

Ross Jr., Frank, *Guided Missiles: Rockets and Torpedos*, New York, Lathrop, Lee & Shepard, 1951.

Weyl, A. R., *Engins téléguidés*, Paris, Dunod, 1952; a translation of *Guided Missiles*, London, Temple Press, 1949.

Gatland, Kenneth W., *Development of the guided missile*, 2nd ed., London, Iliffe, 1954.

Benecke, Th. and A. W. Quick (eds.), *History of German Guided Missile Development*, AGARD First Guided Missile Seminar, Munich, April 1956.

Ordway, Frederick and Ronald C. Wakeford, *International Missile and Spacecraft Guide*, McGraw-Hill, 1960.

Clemow, J., *Missile Guidance*, London, Temple Press, 1962.

Smith, J. R. and A. L. Kay, *German Aircraft of the Second World War*, Putnam, 1972, pp. 645-712.

Spearman, M. Leroy, *Historical Development of Worldwide Guided Missiles*, NASA Technical Memorandum 85658, June 1983.

Benecke, Theodor et al., *Die Deutsche Luftfahrt—Flugkörper und Lenkraketen*, Koblenz, Bernard und Graefe, 1987.

Trenkle, Fritz, *Die Deutschen Funklenkverfahren bis 1945*, Heidelberg, Alfred Hüthig, 1987.

# Chapter 1

# Terminology and Definitions

This chapter is dedicated to definitions of several concepts and terms used in the study of guidance. Some of these definitions are not universal, there being differences between British and American usage as well as between individual authors. We shall use the terminology that seems most accepted, giving alternatives where relevant.

## 1.1 The Three Levels of the Guidance Process

'Guidance' is the process for guiding the path of an object towards a given point, which in general may be moving. If the given point is fixed and the guided object is 'manned' (or is, for example, a migrating bird), then the process is simply *navigation*.[1] Thus, although navigation can be said to be a subclass of guidance, in this book we give the term 'guidance' a narrower meaning, which excludes navigation. This specific meaning will now be explained.

### 1.1.1 Definitions

Guidance is a hierarchical process which may be said to consist of three levels.

(a) In the highest one, a *geometrical rule* is stated in terms of a *line-of-sight* that passes through the objective of the guidance. We shall henceforth refer to this objective as the *target* T, and to the guided object as M. This definition serves to distinguish between guidance in the meaning given to it in this book and marine navigation, for example, or inertial guidance.

---

[1] Sometimes a narrower definition is given for navigation, according to which it is merely the art or science of finding the exact location of an observer relative to earth.

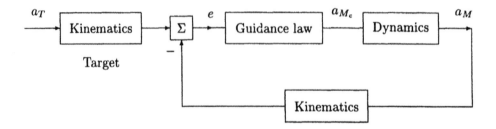

Figure 1.1: The guidance loop

The kinematics of the engagement is dealt with in this level, and problems of shape of trajectories, curvature, and required lateral accelerations are examined. Time $t$, of course, is usually a parameter; very often, the total time of the guidance process, from $t = 0$ when it started till intercept at $t = t_f$ , is of importance. We call it the *time-of-flight* although M may not necessarily be an aircraft or a spacecraft.

**(b)** In the second level of the process, a *guidance law* appears. A guidance law is the algorithm by which the desired geometrical rule is implemented; this is done by closing a *guidance loop* (Fig. 1.1). The '*error*' $e$ in this feedback loop is a direct function of the measured deviation of M's state in space from the state required by the geometrical rule. It is used to produce a *steering command* given to M according to the guidance law.

In most cases, the command is for lateral acceleration, which will be denoted here by $a_{M_c}$, '$c$' signifying 'command'. The lateral acceleration (sometimes called *latax*) actually performed by M is $a_M$. Similarly, $a_T$ denotes the lateral acceleration of T. The resulting motion of M is such that $e$ is reduced by the feedback, eventually driven to zero. .

**(c)** We now come to the third level of the process, that of *control*. Here M is no more just a point. Rather, it is a body, say rigid, with mass and moments of inertia, whose attitude in space is defined by three angles. Clearly, it is in this level that we study the stability and performance of the *body control* loop, which is an inner loop whithin the guidance loop. In some cases, though, there is no body control *loop* per se, as the control is direct, or 'open loop'.

**Note.** Other ways of defining the terms introduced above are encountered in the literature. For example, for aerospace applications of control, Bryson distinguishes between four categories which "often overlap", namely Flight Planning, Navigation, Guidance, and Control. By his definition, 'guidance' is the determination of a *strategy* for following the nominal flight path in the presence of off-nominal conditions, wind disturbances, and navigation uncertainties; "a strategy, of course, involves feedback" [1]. Since feedback is required, this definition is quite close to ours.

### 1.1.2  An Example

For an illustrative example, let us take a three-point guidance system that operates according to the *line-of-sight* (LOS) geometrical rule. By this rule, M must stay on a ray $OT$ from a reference point $O$ to the target $T$, i.e., on the LOS. Note that we use the symbols T and M for the objects themselves and $T$ and $M$ for the points in space that respectively represent them. T is not necessarily stationary. Suppose M can measure its distance from the LOS; then this distance may be regarded as the error $e$ and the guidance law would make $a_{M_c}$ a function of $e$: $a_{M_c} = f(e)$, where $f(.)$ could be linear, saturable, or have some other nonlinear, usually odd-symmetric, characteristics. A more sophisticated law could be $a_{M_c}(t) = k(1 + T_d d/dt)e(t)$, $T_d = const.$, where a derivative term has been added to the linear function.

If M is a missile, then body control is necessary in order that M remain controllable throughout the flight and $a_M$ can follow the commands $a_{M_c}$; for example, so that the response to an 'up' command is indeed 'up' and not 'down' or 'left'.

### 1.1.3  Scope of this Book

In this book we shall deal with the upper two levels of the guidance process hierarchy, namely geometrical rules and guidance laws, and shall not delve into the topic of body control. There are two reasons for this.

The first reason is the wish to make the book as reasonably small as possible, topped by the fact that there are available several good texts on control of aircraft and missiles [2-5] and marine vehicles [6].

The second reason stems from the fact that body-control loops are faster, sometimes much faster, than guidance loops; in control-engineering terminology, body-control loops are high-bandwidth inner loops whithin lower-bandwidth guidance loops (Fig. 1.2). Indeed, in many works it is assumed that the inner loop is so fast that, for the purpose of preliminary outer-loop analysis, it can be ignored altogether. M is then represented in the guidance loop by *zero-order dynamics*, for which $a_M \equiv a_{M_c}$ by definition.

However, when implementation and mechanization of guidance laws are studied, we shall not ignore the fact that body control loops do have dynamics; in some cases, the effects of nonlinearities will also be examined.

## 1.2  Terminology Related to Implementation

### 1.2.1  Definitions

Guidance systems may be *autonomous* or not, depending on whether or not M requires some external aid during the guidance process. Such an aid may for example be

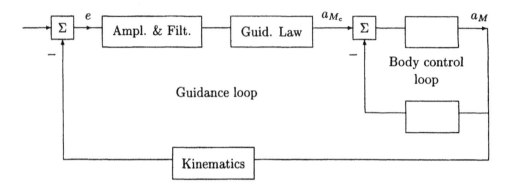

Figure 1.2: Body-control loop whithin the guidance loop

⋆ a reference point in the case of LOS guidance,
⋆ 'illuminating' the target by electromagnetic radiation in the case of certain anti-aircraft systems, or
⋆ creating a laser beam for yet other missile applications.
When such aid is necessary, the system is said to be *non-autonomous*.

*Command systems*, in which commands are transmitted to M (e.g., by radio or by wire) are, of course, non-autonomous. Very often the geometrical rule employed in command systems is LOS guidance, in which M is kept on the LOS *OT*; we then have *command to line-of-sight* (CLOS) guidance. However, there exist command guidance systems which are not CLOS; examples are provided in the next subsection. Conversely, some LOS guidance systems are not command ones: for example, *beam riders*.

Guidance systems that involve M and T only, represented by the points $M$ and $T$, respectively, are called *two-point* systems; when a third point is required, as in LOS guidance, the system becomes *three-point*.

A guidance system is called *homing* if M detects the target T and tracks it thanks to energy emitted by the latter. If the source of the energy is T proper, e.g., radio transmissions, acoustic noise, heat, the homing is called *passive*. If T reflects energy beamed at it by M, then this is *active* homing. In *semiactive* systems, M homes onto energy reflected by T, the latter being 'illuminated' by a source I external to both; see Fig. 1.3.

## 1.2.2   Examples

**(a)** While guiding themselves to catch a prey, hawks perform passive homing. Many kinds of bats use active homing, the illumination being ultrasound pulses.

**(b)** Semiactive homing systems (such as the US Sparrow air-to-air missile system)

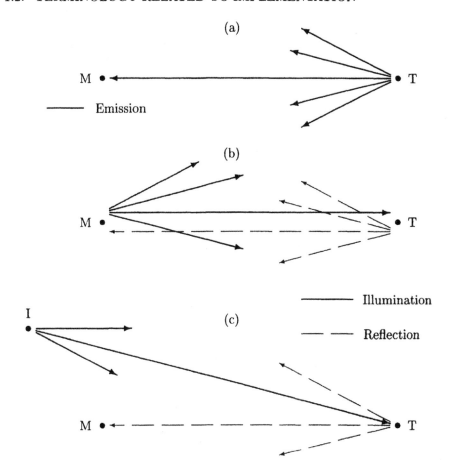

Figure 1.3: (a) Passive, (b) active, (c) semiactive, homing

are non-autonomous, though they do not belong to the command guidance class.

**(c)** The German World War II air-to-sea bomb Hs-293D (which did not become operational) had a TV camera mounted on it and a corresponding radio link, such that the bomb-operator at the aircraft which released the bomb could see the target while the bomb was approaching it. The operator transmitted commands to the bomb according to the location of T on his screen [7]. Hence this bomb was command-guided, though not CLOS.

**(d)** A medium-range weapon may be command-guided during the first part of its flight, and homing during the latter part. In the Israeli Gabriel sea-to-sea missile, guidance during the first part of the fight is CLOS, and during the latter one, semiactive homing *proportional navigation* (PN), to which two chapters of this book are dedicated. In a recent study of a ground-to-air missile system, the same guidance law, namely PN, is used in both parts of the flight, although homing prevails during the latter part only [8].

(e) There exist air-defense systems where M locks on T and tracks it, transmitting measurements to a ground control station. There, the necessary computations are carried out and, based upon them, steering commands are transmitted to M. Such systems are called *track via missile*: they seem to belong to both the command guidance category and the homing one.

   **Note.** Until the early 1950's, 'homing' was synonymous with 'following the pure-pursuit geometrical rule', and 'guided missile' usually meant 'command guided'.

## 1.3   Geometry and Kinematics

### 1.3.1   Basic Definitions and Notations

In this section we introduce several concepts which will be used often in subsequent chapters.

   Let $O$ be the origin of an inertial reference *frame of coordinates* (FOC). The positions of M and T in this FOC are given by the vectors $\mathbf{r}_M \triangleq OM$ and $\mathbf{r}_T \triangleq OT$, respectively. In two-point guidance systems, the vector

$$\mathbf{r} \triangleq \mathbf{r}_T - \mathbf{r}_M \tag{1.1}$$

is conventionally called the *range*. Its time derivative

$$\dot{\mathbf{r}} = \dot{\mathbf{r}}_T - \dot{\mathbf{r}}_M \triangleq \mathbf{v}_T - \mathbf{v}_M \tag{1.2}$$

is the *relative velocity* between the two objects, and $\mathbf{v}_T$ and $\mathbf{v}_M$ are the velocities of T and M, respectively. The *closing velocity*, a term often used in the study of guidance, is simply

$$\mathbf{v}_C \triangleq -\dot{\mathbf{r}}. \tag{1.3}$$

   The ray that starts at $M$ and is directed at $T$ along the positive sense of $\mathbf{r}$ is called the *line of sight* (LOS), a concept which we have already used in the beginning of the chapter and which will be a central pivot in most of our analyses. In three-point guidance systems, where an external reference point $O$ is used, there are actually involved *two* lines of sight, $OM$ and $OT$, along $\mathbf{r}_M$ and $\mathbf{r}_T$, respectively.

   In general, the guidance process takes place in three-dimensional (3-D) space. A brief reminder of the vector equations of motion in 3-D is given in Appendix A, and techniques for rotational transformations of 3-D vectors are shown in Appendix B. Since analysis of motion in space, whether guided or not, is usually complex, most authors tackle guidance problems by first assuming that the engagements are *planar*, i.e., that they take place on a fixed plane. Sometimes this assumption is not unrealistic: for example, if M is a boat, or a car. In other cases the assumption may be an approximation valid for short periods of time. In fact, certain guidance processes evolve into being planar even if they started otherwise.

Thus, a three-point guidance system is said to be *planar* if the points $O$, $T$, and $M$ and the velocities $\mathbf{v}_T$ and $\mathbf{v}_M$ remain in the same fixed plane throughout the engagement. The said points and vectors may be on a nonfixed plane throughout a nonplanar engagement; in such a case the situation is said to be always *instantaneously planar*.

A two-point guidance process is said to be planar if $\mathbf{r}$, $\mathbf{v}_T$ and $\mathbf{v}_M$ remain in the same fixed plane. As in three-point guidance, the said three vectors may be always *instantaneously coplanar* even in nonplanar engagements. For more details on planarity and coplanarity, see Appendix A.

A reminder on the kinematics of planar motion will now be presented.

## 1.3.2  Kinematics of Planar Motion

When studying planar engagements we shall use Cartesian FOC $(x, y)$ or $(x, z)$; angles will be defined positive anticlockwise, whether the third (i.e., missing) axis is $z$ or $y$.

Let P be located at the point $(x, y)$ and have the velocity $\mathbf{v}$. We define $\mathbf{r} \triangleq OP$ although in specific guidance problems $OP$ may represent $\mathbf{r}_T$, say, or $\mathbf{r}_M$. Then (see Fig. 1.4),

$$\mathbf{r} = x\mathbf{1}_x + y\mathbf{1}_y = (r\cos\lambda)\mathbf{1}_x + (r\sin\lambda)\mathbf{1}_y \tag{1.4}$$

and

$$\mathbf{v} = \dot{\mathbf{r}} = v_x\mathbf{1}_x + v_y\mathbf{1}_y , \tag{1.5}$$

where $\mathbf{1}_x$ and $\mathbf{1}_y$ are the unit vectors, $(r, \lambda)$ are the polar coordinates of $P$, and $v_x = \dot{x}, v_y = \dot{y}$.

$\mathbf{v}$ can also be resolved into two components *along* and *across* $\mathbf{r}$, respectively, the former being the *radial velocity* $\mathbf{v}_\parallel$ and the latter, the *tangential velocity* $\mathbf{v}_\perp$. Their values are given by the equations

$$\begin{aligned} v_\parallel &= \dot{r} = v\cos\delta \\ v_\perp &= r\dot{\lambda} = v\sin\delta , \end{aligned} \tag{1.6}$$

where $\delta$ is the angle between $\mathbf{r}$ and $\mathbf{v}$ (Fig. 1.4).

The angle $\gamma = \lambda + \delta$ that $\mathbf{v}$ forms with the $x$ axis is called the *path angle* of P. It is recalled that the slope of the trajectory of P at $(x, y)$ is $\tan\gamma$.

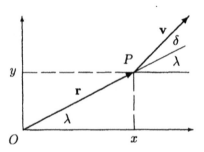

Figure 1.4: Kinematics of planar motion

For constant $v$, it is well known from elementary mechanics that the acceleration $a$ of P is given by the product

$$a = v\dot{\gamma} \ , \tag{1.7}$$

**a** being normal to **v**.

It is often useful to use vector expressions in planar situations as well as in 3-D ones (see details in Appendix A):

**a** can be expressed as the sum of two vectors (see (A.9) in Appendix A), along and across **r**, respectively:

$$\mathbf{a} = \mathbf{a}_{\parallel} + \mathbf{a}_{\perp} = (\ddot{r} - r\omega^2)\mathbf{1}_r + (r\dot{\omega} + 2\omega\dot{r})\mathbf{1}_t \ , \tag{1.8}$$

where $\omega = \dot{\lambda}$, and $\mathbf{1}_r$ and $\mathbf{1}_t$ are the unit vectors along and across **r**, respectively.

It follows from (A.3) that

$$\omega = \frac{\|\mathbf{r} \times \mathbf{v}\|}{r^2} = \frac{xv_y - yv_x}{r^2} \ , \tag{1.9}$$

from (A.13) that

$$\omega = \frac{v \sin \delta}{r} = \frac{v_{\perp}}{r} \ , \tag{1.10}$$

and from (A.6) that

$$\dot{r} = \frac{\mathbf{r} \bullet \mathbf{v}}{r} = \frac{xv_x + yv_y}{r} \ . \tag{1.11}$$

## 1.4   References

[1] Bryson Jr., A. E., "New Concepts in Control Theory, 1959-1984", J. Guidance, Vol. 8, No. 4, 1985, pp. 417-425.

[2] Etkin, Bernard, *Dynamics of Flight - Stability and Control*, 2nd ed., John Wiley, 1982.

[3] Mclean, Donald, *Automatic Flight Control Systems*, Prentice-Hall, 1990.

[4] Blakelock, John, *Automatic Control of Aircraft and Missiles*, 2nd ed., Wiley-Interscience, 1991.

[5] Jenkins, Philip N., "Missile Dynamic Equations for Guidance and Control Modeling and Analysis", *US Army Missile Command, Redstone Arsenal*, Technical Report RG-84-17, April 1984.

[6] Fossen, Thor I., *Guidance and Control of Ocean Vehicles*, John Wiley, 1994.

[7] Münster, Fritz, "A Guiding System Using Television", in Benecke, Th. and A. W. Quick, eds., *History of German Guided Missiles Development*, AGARD 1st Guided Missiles Seminar, Munich, April 1956, pp. 135-160.

[8] Williams, D. E., B. Friedland, and J. Richman, "Integrated Guidance and Control for Combined Command/Homing Guidance", *Proc. Amer. Cont. Conf.*, Atlanta, June 1988, pp. 549-554.

## 14. REFERENCES

# Chapter 2

# Line-of-Sight Guidance

## 2.1 Background and Definitions

For many years, guiding an object M, whether a boat or a missile, meant 'keeping it on the line-of-sight between a reference point and the target', the reference point $O$ usually being near the start point of M. Thus, according to this simple geometrical rule, which will be referred to as the *line-of-sight (LOS) guidance rule*, M is always on the ray that starts at $O$ and passes through $T$, where the target T is located (Fig. 2.1), such that M 'covers' T, as it were. (Recall that we use the symbols T and M for the objects themselves and $T$ and $M$ for the points in space that respectively represent them). Indeed, this rule has been called *Deckung* in German, meaning 'covering'. In general, T is not stationary, and in some cases, neither is O: for example, in ship-defense missile systems.

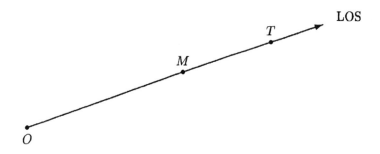

Figure 2.1: LOS-guidance geometrical rule

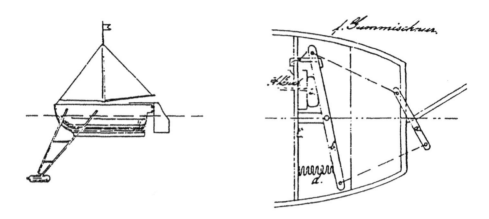

Figure 2.2: Guided boat proposed by Siemens (Source: Reference 2, with kind permission from Springer-Verlag GmbH, Heidelberg)

## 2.2   A Little History

The first guided-weapon proposed in history was supposed to use this geometrical rule, applying teleguidance technique, later called *command*. The system was proposed by Werner Siemens and submitted to the Prussian Ministry of War in August 1870.[1] It consisted of a torpedo mounted beneath a sailing boat, controlled by pneumatic pulses transmitted through rubber tubes (Fig. 2.2). The commands were to be transmitted from a control post on land or on a marine vessel, the position of the guided boat being marked by a shielded lamp. By the time the system had finally been developed and deployed by the German navy—in 1916—the boats were propulsed by advanced internal-combustion engines, could achieve speeds exceeding 30 knots (15 m/sec), and were guided from airborne command posts via radio and 50km-long electrical cables. In October 1917 the first operational success was attained when a British ship was hit and sunk [2].

It seems that the German government followed to the letter the clause of the Versailles treaty (1919) that forbade Germany to *use* guided "pilotless aircraft",

---

[1]Werner von Siemens (1816-1892) is better known for his achievements in telegraphy and in electrical machinery engineering, for having invented the electric locomotive, and for being the cofounder of the well-known Siemens and Halske firm in 1847. In his volume of memoirs, only a few lines are dedicated to the proposal for "steering unmanned boats, furnished with explosives" and to the "politically stirring time" that brought it about [1].

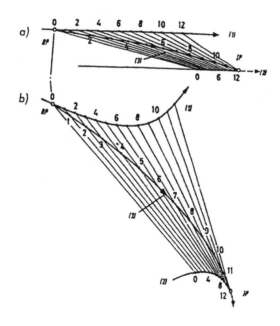

Figure 2.3: Trajectories of the Hs-293 (Source: Reference 9, with kind permission from AGARD/NATO)

but resumed active *studies* of guided-weapon systems, as well as of aeronautics and rocketry, in the late 1920's [3]. Germany soon became, though few suspected it, the world's leader in these fields of military technology.

It is no wonder that in World War II, as of 1943, Germany already had two operational guided-weapon systems, namely the SD-1400-X, nicknamed Fritz X, and the Hs-293 radio-guided air-to-sea bombs. They started their operational life whithin days of each other, in August 1943. Both had impressive success: as a matter of fact, the former was later used for destroying bridges on the Oder, when Russian forces were pushing into Germany on the eastern front [3-7].

All air-to-sea weapons of that epoch used the LOS-guidance geometrical rule, which is the first and best known variant of three-point-guidance rules. An illustration of trajectories involved in the guidance of the Hs-293 is shown in Fig. 2.3, where (a) is the vertical projection, (b) is the horizontal one, RP denotes the release point and IP, the impact point [9]. When M is kept on the LOS by commands transmitted to it, the guidance is nowadays called *command to line-of-sight* or *CLOS*.

The same geometrical rule was used for guiding ground-to-air missiles. Quite a few such systems were developed in Germany, the best known being Wasserfall,

Fueurlilie, Rheintochter, Enzian, and Schmetterling.  None achieved operational status, but some may have been close to it when the war ended in May 1945. In all of these systems, command was by radio link; in some of them, targets were to be tracked by optical as well as radar devices.

This is not the whole story of CLOS guidance in World War II. An *air-to-air* missile, the X-4, which started full-scale production but did not see actual combat, was command-guided by wire according to that rule, too [5, 6]. This may seem odd, but a US Patent dated March 1979 again proposes a CLOS-guided air-to-air missile system, the commands now to be transmitted by radio [10].

The X-4 had an *anti-tank* derivative, the X-7, called Rotkäppchen ('Red Riding Hood'). It was operational towards the end of the war on a very minor scale and became the predecessor of several CLOS-guided anti-tank missiles developed by both the Western and the Eastern powers in the early 1950's.

For more information on the history of guided missile development, in Germany in particular, the reader is referred to the literature cited above [3-7]; two of these references, the more recent ones, provide authoritative references to the literature [6,7].

## 2.3   Kinematics

In this section we shall examine the kinematics of trajectories produced when the LOS-guidance geometrical rule is applied. We start with the planar case.

### 2.3.1   Analysis of the Planar Case

(a) We recall that in the planar case, the points $O$, $T$ and $M$ and the velocity vectors $\mathbf{v}_T$ and $\mathbf{v}_M$ are on the same fixed plane (Fig. 2.4). $Ox$ represents a reference line, from which the *LOS angle* $\lambda$ is measured; from the same reference, T's *path angle* $\gamma_T$ and M's *path angle* $\gamma_M$ are measured. $\mathbf{v}_T$ and $\mathbf{v}_M$ form the angles $\theta$ and $\delta$ with the LOS, respectively. The ranges $\mathbf{r}_T$ and $\mathbf{r}_M$ are $OT$ and $OM$, respectively.

Since $M$ and $T$ are on the same ray, they have the same angular velocity $\dot{\lambda}$. From this property and (1.10) we derive the equations

$$\dot{\lambda} = \frac{v_T \sin\theta}{r_T} = \frac{v_M \sin\delta}{r_M} \; , \tag{2.1}$$

which result in the equation

$$\sin\delta = \frac{1}{K} \frac{r_M}{r_T} \sin\theta \tag{2.2}$$

where K is the velocity ratio

$$K \triangleq \frac{v_M}{v_T} \; . \tag{2.3}$$

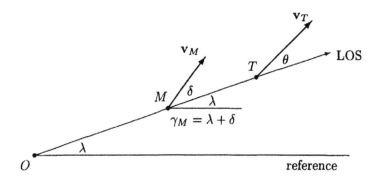

Figure 2.4: Geometry of planar LOS guidance

It follows from (2.2) that, no matter what T's maneuver (i.e., $\dot\gamma_T$) or $K$ are (provided $K > 1$),

(i) $\sin\delta$ has the same sign as $\sin\theta$; in other words, $\mathbf{v}_M$ always leads $\mathbf{r}_M$ (this is why $\delta$ is called the *lead angle*);

(ii) M starts its flight, when $r_M/r_T = 0$, having zero lead-angle $\delta$;

(iii) Towards capture, or interception, when $r_M/r_T = 1$, there exist *collision-course conditions*, in which, by definition, $\sin\delta = (1/K)\sin\theta$, or

$$v_M \sin\delta = v_T \sin\theta .$$

We shall elaborate on these conditions when we discuss parallel navigation in Chapter 4.

(b) We now recall (see Sec. 1.3.2) that $\mathbf{v}_M$ is conveniently resolved into two components, along and across the LOS, $\mathbf{v}_{\|}$ and $\mathbf{v}_{\perp}$, respectively, given by the equations

$$\begin{cases} v_{\|} = \dot r_M = v_M \cos\delta \\ v_{\perp} = r_M \dot\lambda = v_M \sin\delta , \end{cases} \tag{2.4}$$

from which it follows that that

$$\left(\frac{dr_M}{d\lambda}\right)^2 + r_M^2 = \left(\frac{v_M}{\dot\lambda}\right)^2 . \tag{2.5}$$

When $\dot\lambda(\lambda)$ is known, the trajectory of M in the polar coordinates $(r_M, \lambda)$ can in principle be obtained from (2.5). A few examples will be shown in Sec. 2.3.2; in general, however, one has to resort to numerical calculations.

(c) Let us now examine the lateral acceleration $a_M$ (also called *maneuver* and *latax*) required by M in order to stay on the LOS. By the equation $a_M = v_M \dot\gamma_M$

(see Sec. 1.3.2) and the identity $\gamma_M = \lambda + \delta$ (see Fig. 2.4) we get from (2.1) and (2.4) the following expressions.

$$a_M = 2v_M \dot{\lambda} + \frac{r_M \ddot{\lambda}}{\cos \delta} - \dot{v}_M \tan \delta = 2v_M (\dot{\lambda} + \frac{r_M (\ddot{\lambda} - \dot{v}_M \dot{\lambda}/v_M)}{2\dot{r}_M}). \qquad (2.6)$$

If $v_M$ is constant, or at least if $\dot{v}_M$ is small compared to the ratio $v_M \ddot{\lambda}/\dot{\lambda}$, which is very often the case, (2.6) simplifies to the compact expression

$$a_M = 2v_M (\dot{\lambda} + \frac{r_M \ddot{\lambda}}{2\dot{r}_M}) . \qquad (2.7)$$

From this equation the required maneuver is seen to equal the sum of two terms, the first one being proportional to $\dot{\lambda}$ and the second one, to $\ddot{\lambda}$ weighted by the factor $r_M/2\dot{r}_M$. Thus, when guidance begins, $a_M$ is determined by $\dot{\lambda}$; as it proceeds, the contribution of $\ddot{\lambda}$ increases, roughly in proportion to $r_M$.

**Note.** An interesting property of LOS guidance is that, when M is still near the origin $O$,

$$a_M \approx 2v_M \dot{\lambda} ,$$

which is twice the value that would be required if M were guided by another geometrical rule, namely *pure pursuit*, to be discussed in Chapter 3.

### 2.3.2   Examples for the Planar Case

**Example 1 - Nonmaneuvering T**

**(a)** Let our plane be $z = 0$ in a Cartesian FOC, and let T be travelling according to the equations

$$x(t) = x_0 - v_T t, \quad y(t) = h ,$$

$v_T$ and $h$ being constant. This could represent an air-defense engagement, where $z = 0$ would be a fixed vertical plane and T, an in-coming aircraft flying at the constant altitude $h$. A missile M is guided from the origin 0 to intercept T. The expressions for $\dot{\lambda}$ and $\ddot{\lambda}$ are easily obtained, as follows.

$$\begin{cases} \dot{\lambda} = \frac{v_T \sin \lambda}{r_T} = \frac{v_T}{h} \sin^2 \lambda \\ \ddot{\lambda} = 2(\frac{v_T}{h})^2 \sin^3 \lambda \cos \lambda . \end{cases} \qquad (2.8)$$

(Note that $\dot{\lambda}$ has a maximum at $\lambda = 90°$ and $\ddot{\lambda}$ at $\lambda = 60°$). Hence, by (2.5), the differential equation for the present case is found to be

$$(\frac{dr_M}{d\lambda})^2 + r_M^2 = \frac{(Kh)^2}{\sin^4 \lambda} . \qquad (2.9)$$

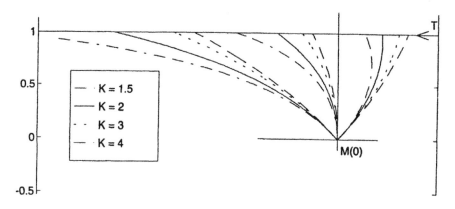

Figure 2.5: Trajectories for $\lambda_0 = 45^o$, $90^o$, and $135^o$

This equation has been studied as early as in 1931, in connection with the following problem: "An automobile moves along a straight road with a constant speed ($v$) while a man in a field beside the road walks with a constant speed ($u$) along such a path as to always keep a tree between him and the automobile. Determine his path" [8]. In this scenario, M and T are on opposite sides of $O$ (the tree) rather than on the same side, as in LOS guidance.

(b) Before general solutions to (2.9) are discussed, several trajectories obtained by digital simulation will be shown for an illustration in Figs. 2.5 and 2.6.

In the first one, T moves from right to left and three salvos, as it were, are launched to intercept it, at the points in time when $\lambda = 45^o$, $90^o$, and $135^o$, respectively. Each salvo consists of four M's, having the relative velocities $K = 1.5, 2, 3$, and 4, respectively. See Fig. 2.5.

It is often of interest to examine the trajectories of M *from the point of view of* T, i.e., in a frame of coordinates (FOC) attached to T. For example, when one wishes to compare the present three-point geometrical rule with two-point ones. Such trajectories are referred to as *relative*, and the location of M in this FOC is usually expressed in polar coordinates *range r* and *aspect angle* $\psi$. We will adopt the convention according to which the reference axis of this FOC is the direction of $-\mathbf{v}_T$. In other words, $\psi$ is measured from the *target's tail*, and hence the abbreviation we shall often use is TT FOC (target's tail frame of coordinates).

In Fig. 2.6, trajectories in the TT FOC are shown for the same guidance cases illustrated in Fig. 2.5. The target T faces "up", and the A, B, and C salvos are for $\lambda_0 = 45^o$, $90^o$, and $135^o$, respectively, i.e., when $\psi_0 = -135^o$, $-90^o$, and $-45^o$, respectively.

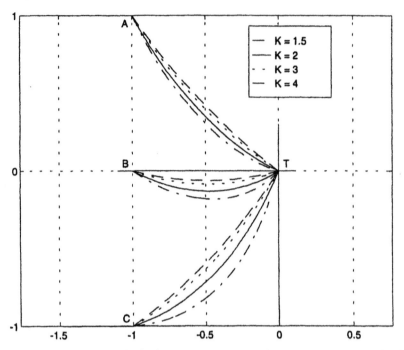

Figure 2.6: Relative trajectories for the guidance cases of Fig. 2.5

(c) We now introduce dimensionless variables for displacements, time, and acceleration, as follows:

$$
\begin{cases}
r_M^* \triangleq \frac{r_M}{Kh}, \ x^* \triangleq \frac{x_M}{Kh}, \ y^* \triangleq \frac{y_M}{Kh} \\
\qquad\quad t^* \triangleq \frac{v_T}{h}t \\
\qquad\quad a_M^* \triangleq \frac{Kh}{v_M^2}a_M \ .
\end{cases}
\tag{2.10}
$$

Using these newly-defined variables we get from (2.9) the nondimensional differential equation

$$
(\frac{dr_M^*}{d\lambda})^2 + r_M^{*2} = \frac{1}{\sin^4\lambda} \ .
\tag{2.11}
$$

This equation can be solved in terms of elliptic integrals or, of course, numerically (see Appendix D.1). Some results are presented in the $(x^*, y^*)$ plane, Fig. 2.7, where the (dimensionless) 'altitude' of T is, by (2.10), $y^* = h/Kh = 1/K$. Three graph-families are shown, as follows.

Heavy lines represent trajectories, all starting at the origin. Along them, $t^*$ is shown as a parameter, dots placed at $t^*$-intervals of 0.1. *Isochrones*, i.e., lines

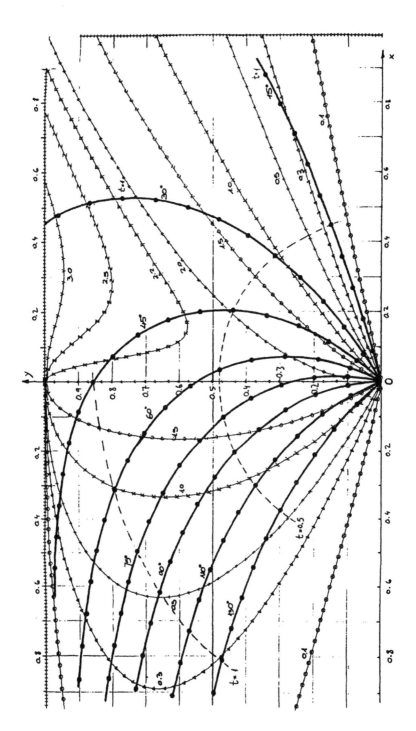

Figure 2.7: The Kh plane, Example 1

of equal time, are shown for $t^* = 0.5$ and $1.0$. (Note that the isochrones may be roughly approximated by circles with radii $t^*$). Thinner lines are *isomaneuver* lines, i.e., lines of equal $a_M^*$. We shall refer to the coordinates and graphs shown in Fig. 2.7 as the *Kh plane*.

For a numerical illustration, suppose $v_T = 300 \ m/sec$, $v_M = 600 \ m/sec$, $h = 3000 \ m$, and $\lambda_0 = 60^\circ$. Thus, referring to the Kh plane, T flies on the $y^* = 1/K = 300/600 = 0.5$ line. Upon examining the graphs of Fig. 2.7 it is seen that M intercepts T at $t^* \approx 0.52$, when $\lambda \approx 87^\circ$, i.e., almost directly above $O$. Maximal lateral acceleration is required about then, its value being close to 2 (the line $x^* = 0$ being an $a_M^* = 2$ line). This value is translated into $2 \times 600^2/(2 \times 3000) = 120 \ m/sec^2$, or about 12 g. ($g = 9.8 \ m/sec^2$, the acceleration of gravity, a unit of acceleration often used by aeronautical engineers).

*Note.* Although obtaining the trajectories and isochrones for the Kh plane, Fig. 2.7, requires some numerical effort, obtaining approximate equal-$a_M^*$ lines for the same plane is straightforward. From (2.4), (2.7), (2.8), and (2.10) we get the expression

$$a_M^* = 2\sin^2 \lambda + \frac{2r_M^* \cos \lambda \sin^3 \lambda}{\sqrt{1 - r_M^{*2} \sin^4 \lambda}} , \qquad (2.12)$$

from which, by solving for $r_M^*$, the equation for an equal-$a_M^*$ line is obtained, as follows:

$$r_M^*(\lambda) = \left[ (\frac{\sin^3 \lambda \cos \lambda}{a_M^*/2 - \sin^2 \lambda})^2 + \sin^4 \lambda \right]^{-\frac{1}{2}} .$$

The $y^*$-axis is a line for $a_M^* = 2$: this of course is whenever T is directly 'above' $O$. Another $a_M^* = 2$ line is the parabola $x^* = y^{*2}$. Also note that when M approaches the $y^* = 1$ line, $a_M^* \to 4\sin^2 \lambda$.

(d) Very often, the *interception zone* for a LOS guidance system is determined by the limitation on the lateral acceleration M can perform. Suppose the maximum value is $a_{M_{max}}$, and $a_M$ attains it precisely at intercept, where $y_M = y_{M_f} = h$ and $\lambda = \lambda_f$ (this is usually the case when $\lambda_f < 60^\circ$, say). Then, by (2.12) and (2.10),

$$a_{M_{max}} = \frac{v_M^2}{Kh} \left( 2\sin^2 \lambda_f + \frac{2(r_{M_f}/Kh)\cos \lambda_f \sin^3 \lambda_f}{\sqrt{1 - (r_{M_f}/Kh)^2 \sin^4 \lambda_f}} \right) \qquad (2.13)$$

where $a_{M_{max}} = (v_M^2/Kh)a^*{}_{M_{max}}$ and $r_{M_f}/Kh$ has been substituted for $r^*{}_{M_f}$. Since $r_{M_f} \sin \lambda_f = h$ by the geometry, this equation simplifies into

$$a_{M_{max}} = \frac{v_M^2}{Kh} 2\sin^2 \lambda_f \left( 1 + \frac{\cos \lambda_f}{\sqrt{K^2 - \sin^2 \lambda_f}} \right) . \qquad (2.14)$$

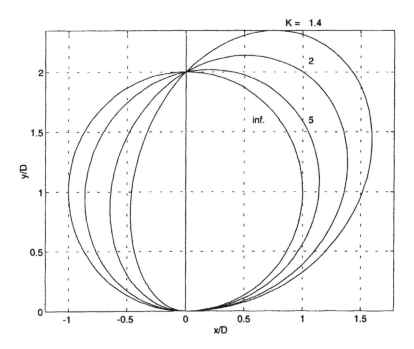

Figure 2.8: Interception boundaries for Example 1 (d)

Using this equation, the boundary of the interception zone can be drawn. We use here normalized coordinates $x_M/D$, $y_M/D$, where

$$D \triangleq v_M^2/Ka_{M_{max}} = v_T v_M/a_{M_{max}} .$$

Four boundaries are shown for an illustration in Fig. 2.8, for $K = 1.4$, 2, 5, and $\to \infty$ (when the boundary turns out to be a circle).

Let $v_T = 200 \, m/sec$, $v_M = 400 \, m/sec$, and $a_{M_{max}} = 100 \, m/sec^2$. Then, since $K = 400/200 = 2$ and $D = 200 \times 400/100 = 800 \, m$, one finds from Fig. 2.8 that the interception zone is bounded in the $y$ direction by $2.15 \times 800 = 1720 \, m$. If T flies at a lower altitude than $1720 \, m$, say $2.0 \times 800 = 1600 \, m$, intercept must occur before $x_T = 1.0 \times 800 = 800 \, m$ (i.e., to the right of that point) in order that $a_M$ does not saturate. (Note that, since $D$ is inversely proportional to $a_{M_{max}}$, the interception zone would be better (i.e., smaller) if $a_{M max}$ were higher).

In some cases, the limitation is on the lead angle $\delta$ (e.g., because limited beam-width beacons or antennas are mounted on M, which must be observable from O). In practice, this limitation is significant only when $K$ is low, since $\sin \delta \leq 1/K$ by (2.2).

**Example 2 - T Moving on a Circle**

In the present example, T executes a constant-speed, constant-maneuver planar motion, say in the plane $z = 0$, such that it is moving on a circle with radius $c = v_T^2/a_T$. Without loss of generality, T's trajectory can be described by the parametric equations

$$x_T(t) = c\cos\omega t, \ y_T(t) = c\sin\omega t \ ,$$

$\omega$ being given by

$$\omega = \frac{v_T}{c} = \frac{a_T}{v_T} \ .$$

Suppose M is launched at $t = 0$ from the origin $O$ to intercept T, and LOS guidance is maintained all the way, $O$ being the reference point. $v_M$ is also assumed constant such that $K = v_M/v_T \geq 1$. It follows from the geometry of the engagement that

$$\dot{\lambda} = \omega.$$

Strangely enough, it turns out when (2.5) is now solved that M's trajectory is circular, too, described by the polar equation

$$r_M(\lambda) = Kc\sin\lambda \ , \tag{2.15}$$

or, equivalently, by the parametric equations

$$x_M(t) = \frac{Kc}{2}\sin 2\omega t, \ y_M(t) = \frac{Kc}{2}(1 - \cos 2\omega t).$$

The center of the circle is at $(0, Kc/2)$ and the radius is $Kc/2$ . M's lateral acceleration $a_M$ equals $2Ka_T$. In Fig. 2.9, the trajectories of T and M are shown for $K = 1, 2$, and 4.

If the reference point is at $(-c, 0)$ rather than $(0, 0)$, then M still moves along a circle, now centered at $(-c, Kc)$ and having the radius $Kc$, where $K$ may now be smaller than 1. The ratio $a_M/a_T$ decreases from $2K$ to $K$. For this case, trajectories are shown in Fig. 2.9 for $K = 0.5, 1$, and 2. (Note that for $K = 0$, i.e., $v_M = 0$, M also scores a hit while staying at the reference point...).

**Example 3 - Free-falling T**

In this example, an analytic solution of (2.5) is not possible, and one has to resort to a numerical one.

Suppose T starts a free fall at $t = 0$ from the point $(0, h)$ having the initial velocity $v_{T_0}$ in the $x$ direction. The $y$ axis is vertical such that the acceleration of gravity $g$ is in the $-y$ direction. As in Example 1, M is LOS-guided from the origin

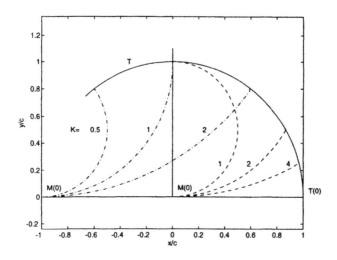

Figure 2.9: Trajectories, Example 2

$(0,0)$ to intercept T. Its velocity $v_M$ is constant, the *initial* velocity ratio being $K_0 = v_M/v_{T_0}$.

From elementary mechanics it is known that, provided air-resistance effects are neglected, the trajectory of T is the parabola given by the parametric equations

$$x_T(t) = v_{T_0}t, \quad y_T(t) = h - gt^2/2 \ .$$

For the present example, $v_{T_0}$ is chosen to be $\sqrt{2gh}$; if, furthermore, $h$ and $\sqrt{h/g}$ are chosen as the units for displacement and time, respectively, one gets the nondimensional equations

$$x_T = \sqrt{2}t, \quad y_T = 1 - t^2/2 \ .$$

Fig. 2.10 shows the solution for several values of $K_0$. There is a minimum value for $K_0$ required for intercept, which turns out to be about 1.29. However, accelerations $a_M$ increase rapidly with $K_0$, while lead angles $\delta$ decrease. (The former effect is due to the relation $a_M = v_M\dot{\gamma}_M = Kv_T\dot{\gamma}_M$; the latter results from the fact that $\sin\delta$ is inversely proportional to $K$, cf. (2.2)).

## 2.3.3 Remarks Concerning Practical Applications

Objects guided according to the LOS geometrical rule tend to require high lateral accelerations in fast-dynamics situations, i.e., when the LOS angle has high rates and high accelerations. This is evident from (2.7) or the more general (2.6). The

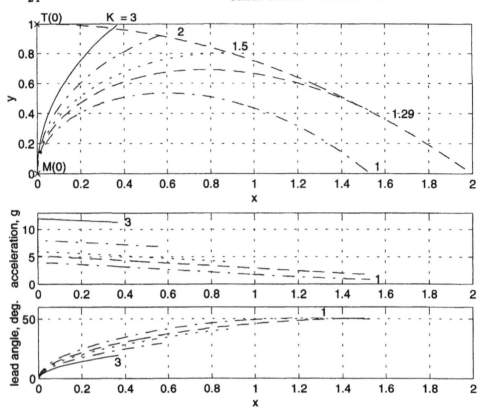

Figure 2.10: Trajectories, accelerations $a_M$, and lead angles $\delta$, Example 3

examples shown in Sec. 2.3.2 illustrate this property. It is therefore not surprising that LOS guidance is dominant in anti-tank missile systems and in anti-aircraft systems that deal with high-flying targets: in both cases $\lambda$ and $\dot{\lambda}$ are low.

From the aeronautical engineering (or marine engineering, if M is a marine vessel) point of view, the lower the required $a_M$, the better. Indeed, the wish to reduce lateral acceleration requirements has produced a *modified* geometrical rule for this type of guidance, with which we shall briefly deal in Sec. 2.3.5.

Another practical difficulty inherent to LOS guidance, not immediately obvious from the analyses and examples given above, results from the limited accuracy with which $\lambda$ and its derivatives can be measured by the system trackers. Since the noise involved with these measurements effectively increases with the range of the tracked objects, accuracy decreases with range. This is why LOS guidance is limited in practice to relatively close-range applications and to weapon systems in which large warheads compensate for low accuracy at high ranges.

### 2.3.4  Kinematics in 3-D Vector Terms

Defined for the general, not necessarily planar, engagement, the LOS guidance geometrical rule is neatly stated by the vector product

$$\mathbf{r}_M \times \mathbf{r}_T = \mathbf{0} \tag{2.16}$$

with the scalar product $\mathbf{r}_M \bullet \mathbf{r}_T > 0$ (otherwise the ray $OM$ would point *away* from $T$ rather than *at* $T$). It follows from (2.16) and the definition of the geometrical rule that

$$\frac{\mathbf{r}_M}{r_M} = \frac{\mathbf{r}_T}{r_T} = \mathbf{1}_r$$

and that $\mathbf{w}_M$ and $\mathbf{w}_T$, the angular velocities of $\mathbf{r}_M$ and $\mathbf{r}_T$, respectively, equal each other. Hence (see Appendix A),

$$\frac{\mathbf{r}_M \times \mathbf{v}_M}{r_M^2} = \frac{\mathbf{r}_T \times \mathbf{v}_T}{r_T^2} \ ,$$

or

$$\mathbf{1}_r \times \mathbf{v}_M = (\frac{r_M}{r_T})\mathbf{1}_r \times \mathbf{v}_T \ . \tag{2.17}$$

In planar terms, this equation was stated as (2.1).

Multiplying (a scalar product) (2.17) by either $\mathbf{v}_T$ or $\mathbf{v}_M$ and applying criterion (A.10) of Appendix A, it is seen that $\mathbf{1}_r$, $\mathbf{v}_T$ and $\mathbf{v}_M$ *are coplanar*, although the engagement is not necessarily planar. In other words, guidance according to the present geometrical rule is always *instantaneously* planar. If $\mathbf{v}_T$ is constant and the reference point 0 is stationary, then the engagement is planar, i.e., takes place on a fixed plane.

### Example 1 - T moving on a Circle, M on a Cone

Suppose T is moving circularly on the (say, horizontal) plane $z = h = const$ such that the center of the circle is on the $z$-axis and the radius is $c$. M is LOS-guided from the origin $O$, and the velocities $v_T$ and $v_M$ are constant with $K = v_M/v_T > 1$.

Clearly, in this special case M remains on the cone whose apex is at the origin, its axis of symmetry is the $z$-axis, and its half-angle $\beta$ is $\arctan(c/h) = \arcsin(c/r_T)$. The trajectory that M traces on the cone is what mechanical engineers might call "variable-pitch conical screw thread".

By the geometry of the engagement, the LOS $\mathbf{1}_r$ is on a generatrix of the cone. $\mathbf{v}_T$ is perpendicular to $\mathbf{1}_r$ since it is tangent to the directrix of the cone. Hence, by (2.17),

$$v_M sin\delta = \frac{r_M}{r_T}v_T = \frac{z_M}{h}v_T \tag{2.18}$$

(see Fig. 2.11), where $z_M$ is the component of $\mathbf{r}_M$ along the $z$-axis. (We can derive (2.18) directly from the equation $\mathbf{w}_M = \mathbf{w}_T$). The component of $\mathbf{v}_M$ along the

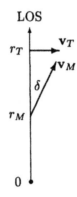

Figure 2.11: Illustration for Example 1, Sec. 2.3.4

same axis is $v_M \cos \delta \cos \beta = \dot{z}_M$. Since $z_M = Kh \sin \delta$ by (2.18), it follows by simple integration that

$$\delta = \frac{v_T \cos \beta}{h}t = \frac{v_T}{r_T}t$$

or

$$t = \frac{r_T}{v_T}\delta = \frac{r_T}{v_T} \arcsin \frac{z_M}{Kh} \, , \qquad (2.19)$$

where $t = 0$ at $z_M = 0$. The engagement ends when $z_M = h$, therefore total engagement time $t_f$ is

$$t_f = \frac{r_T}{v_T} \arcsin \frac{1}{K} \, . \qquad (2.20)$$

At the end of the engagement, i.e., when $t = t_f$, the lead angle $\delta$ equals $\delta_f = \arcsin(1/K)$ and T has traversed the angle $(v_T/c)t_f = (1/\sin \beta)\delta_f$.

The acceleration $a_M$ can be expressed in terms of $a_T$ and $r_M/r_T = z_M/h$, as follows:

$$\frac{a_M}{a_T} = 2K \sin \beta \sqrt{1 + (\frac{r_M \cot \beta}{2K r_T})^2} \, . \qquad (2.21)$$

(See Appendix D.2 for the derivation of this result). Since the value of $K$ is generally high, the present example illustrates, as did the examples in Sec. 2.3.2, the tendency of LOS guidance to require high lateral accelerations, especially at the final stage of the engagement.

A three-dimensional illustration is given in Fig. 2.12. For the case depicted, $h = 4000\,m$, $v_T = 200\,m/s$, $K = 1.2$, $c = 800\,m$, such that $a_T = 50\,m/s^2$. The start point and end point of T's trajectory are marked by "+" and "*" symbols, respectively.

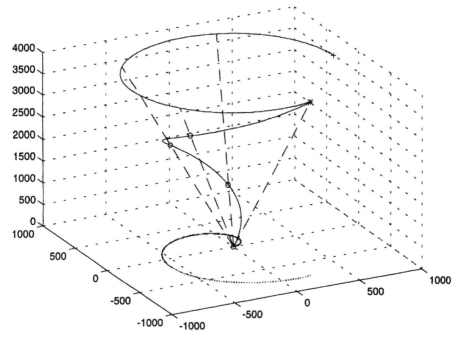

Figure 2.12: Illustration for Example 1, Sec. 2.3.4

## Example 2 - Ground Tracker Kinematics

Tracking the target T is obviously a must for all LOS guidance systems; in CLOS (command-to-LOS) systems, tracking M as well. If the LOS is more or less station-ary, such that T is inside the *seeker* field of view, tracking is accomplished whithin the seeker itself by electronic or manual means (more will be said about seekers and trackers in Sec. 2.5.2). However, when the LOS angles may vary considerably during the engagement, the seeker is usually mounted on a platform which tracks T (or M) in azimuth, i.e., about the nominally-vertical $z$-axis, by convention, and in elevation (about an axis normal to both the $z$-axis and $\mathbf{r}$). In this example we wish to examine some of the kinematics aspects of this kind of tracking. Other platform configurations are possible, of course.

We require of course two rates, namely azimuth rate $\dot{\psi}$ and elevation rate $\dot{\theta}$ (Fig. 2.13). In the present example we will show how these angular rates depend on the target's location and speed.

The tracker is assumed to be at the origin of a Cartesian FOC, from which $\mathbf{r}$ is measured. We can start by finding the LOS rate $\mathbf{w}$. Since $\mathbf{w} = (\mathbf{r} \times \mathbf{v})/r^2$, where $\mathbf{r}$

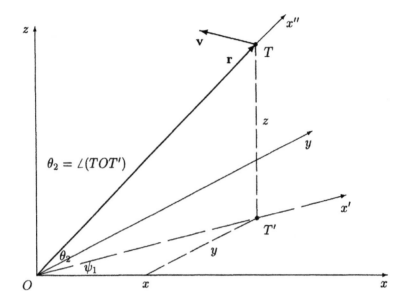

Figure 2.13: Illustration for Example 2, Sec. 2.3.4

is the range to T and $\mathbf{v}$ is the velocity of T, it follows that

$$\mathbf{w} = \frac{1}{r^2} \begin{bmatrix} yv_z - zv_y \\ zv_x - xv_z \\ xv_y - yv_x \end{bmatrix}, \tag{2.22}$$

where $x, y, z$ are the coordinates of $T$.

Let us now define the *seeker FOC*: Its x-axis coincides with $\mathbf{r}$, y coincides with the elevation axis, and z completes a right-angled Cartesin FOC. In order to express $\mathbf{w}$ in this FOC, we rotate the original FOC twice; first by the angle $\psi_1$ about the z-axis, then by the angle $\theta_2$ about the (new) y axis (Fig. 2.13; for details on such rotational transformations, see Appendix B). If these angles are chosen according to the equations

$$\psi_1 = \arctan \frac{y}{x}, \; \theta_2 = -\arcsin \frac{z}{r}, \tag{2.23}$$

then $\mathbf{w}$ in the seeker FOC is given by the equation

$$\mathbf{w} = \frac{1}{r^2} \begin{pmatrix} c\theta & 0 & -s\theta \\ 0 & 1 & 0 \\ s\theta & 0 & c\theta \end{pmatrix} \begin{pmatrix} c\psi & s\psi & 0 \\ -s\psi & c\psi & 0 \\ 0 & 0 & 1 \end{pmatrix} \begin{bmatrix} yv_z - zv_y \\ zv_x - xv_z \\ xv_y - yv_x \end{bmatrix}, \tag{2.24}$$

more concisely written as

$$\mathbf{w} = \frac{1}{r^2} \begin{pmatrix} c\theta c\psi & c\theta s\psi & -s\theta \\ -s\psi & c\psi & 0 \\ s\theta c\psi & s\theta s\psi & c\theta \end{pmatrix} \begin{bmatrix} yv_z - zv_y \\ zv_x - xv_z \\ xv_y - yv_x \end{bmatrix}, \tag{2.25}$$

where, for the sake of brevity, $c(.) = \cos(.)$, $s(.) = \sin(.)$, $\psi = \psi_1$, and $\theta = \theta_2$. The reason for the minus sign in the definition of $\theta_2$ is that elevation angles of seekers, guns, etc. are conventionally measured 'up', i.e., in the *negative* sense about the $y$-axis.

$\dot{\theta}$ is the component of $\mathbf{w}$ along the $y$-axis of the seeker FOC, easily found from (2.25). $\dot{\psi}$ is the component of $\mathbf{w}$ in the $z$-axis divided by $\cos\theta$, since the tracker azimuth rotation is about an axis that forms the angle $\theta$ with the $z$-axis of the tracker FOC. Therefore we finally have the tracker rates

$$\dot{\theta} = \frac{1}{r^2}[-(yv_z - zv_y)\sin\psi + (zv_x - xv_z)\cos\psi] \tag{2.26}$$

and

$$\dot{\psi} = \frac{1}{r^2 \cos\theta}[(yv_z - zv_y)s\theta c\psi + (zv_x - xv_z)s\theta s\psi + (xv_y - yv_x)c\theta] . \tag{2.27}$$

Suppose now, for an illustration, that an aircraft T is approaching an air-defense system, such that it is flying at a constant altitude $z = h$ with a speed $U$ in the $-x$ direction, its $y$ componenet being $c = const$. Then, by (2.26) and (2.27), the rates required for tracking it are

$$\dot{\theta} = \frac{U}{r^2}(-z\cos\psi) = \frac{U}{r}\sin\theta\cos\psi \tag{2.28}$$

and

$$\ddot{\psi} = \frac{U}{r^2 \cos\theta}(-z\sin\theta\sin\psi + y\cos\theta) = \frac{U}{r}\frac{\sin\psi}{\cos\theta} = \frac{U}{c}\sin^2\psi . \tag{2.29}$$

$r$, $\psi$ and $\theta$ are functions of time, since $x = x_0 - Ut$:

$$r = \sqrt{x^2 + c^2 + h^2}, \quad \psi = \arctan\frac{c}{x}, \quad \theta = -\arcsin\frac{h}{r}. \tag{2.30}$$

Note the singularity at $\theta = 90^o$, which may occur if $c = 0$.

## 2.3.5 Kinematics of Modified LOS Guidance

The main motivation for modifying the straighforward LOS guidance has been shown above to be the desire to reduce lateral acceleration requirements.

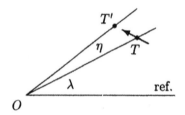

Figure 2.14: Geometry of modified LOS guidance

(a) In a well-known modification, M is required to be on a ray $OT'$ rather than on the LOS $OT$, such that $OT'$ *leads* $OT$ by a certain angle $\eta$ (Fig. 2.14). This geometrical rule is not new: in fact, it has been alluded to in Soviet Union literature in the early sixties, so that presumably it may have been used in such veteran ground-to-air missile systems as the SA-2 or SA-3 (NATO notation) [11]. We present here a brief analysis of a planar engagement [12].

The angle $\eta$ by which $OT'$ leads $OT$ is made to be proportional to $\dot{\lambda}$:

$$\eta(t) = \varepsilon \dot{\lambda} f(t), \quad f(t_f) = 0 , \tag{2.31}$$

where $\varepsilon$ is a constant and $t_f$ is the total time of flight. Note that if $f(t_f)$ were not required to be zero, all missiles would miss the target. A convenient choice for $f(.)$ is to let it be the *time-to-go* $\tau$, i.e.,

$$f(t) = \tau \triangleq t_f - t . \tag{2.32}$$

Time-to-go is a very important concept in guidance, which will be used frequently in subsequent chapters.

More generally, the function $\eta(t)$ is chosen according to the application. For example, if M is an air-to-surface guided bomb, this function may be chosen such that the trajectory of M differs as little as possible from a free-fall one. On the other hand, if T is not stationary, $\eta(t)$ is chosen such that the engagement approaches parallel navigation (see Chapter 4) [11]. In either case, $\eta(t_f)$ must be 0 for the reason explained above.

During the guided flight, $\lambda(t)$ and $\dot{\lambda}(t)$ are measured by the system trackers, $t_f$ is continuously estimated, and $\eta(t)$ is calculated. M is commanded to be on the ray $OT'$ that forms the angle $\mu$ with the reference axis such that

$$\mu(t) = \lambda(t) + \eta(t) . \tag{2.33}$$

We shall now find the kinematic equations for the lateral acceleration. Substituting $\mu$ for $\lambda$ in (2.7), we have an approximate expression for $a_M$, as follows:

$$a_M = 2v_M \left( \dot{\mu} + \frac{r_M \ddot{\mu}}{2 \dot{r}_M} \right). \tag{2.34}$$

From (2.31)-(2.33), the following expressions for $\dot{\mu}$ and $\ddot{\mu}$ are easily obtained.

$$\dot{\mu} = \dot{\lambda} + \varepsilon(\dot{\lambda}\dot{\tau} + \ddot{\lambda}\tau) \tag{2.35}$$

$$\ddot{\mu} = \ddot{\lambda} + \varepsilon(\dot{\lambda}\ddot{\tau} + 2\ddot{\lambda}\dot{\tau} + \lambda^{(3)}\tau) \ . \tag{2.36}$$

Numerical solutions to (2.34)-(2.36) for the engagement of Example 1, Sec. 2.3.2, have been found in terms of the Kh plane [12]. Results for $\varepsilon = 0.3$ are shown in Fig. 2.15, which is completely analogous to Fig. 2.7, drawn in Sec. 2.3.2 for the regular (i.e., $\varepsilon = 0$) case. Comparing the $\varepsilon = 0.3$ case to the regular one, we can immediately make two observations. First, M's trajectories, while changed, retain their general form; second, the equal-$a_M^*$ lines are pushed up, towards the $y^* = 1$ line. For example, for a target flying at the (dimensionless) altitude $y^* = 0.9$, maximum $a_M^*$ found from Fig. 2.15 is less than 2.0, compared to about 2.8 found from Fig. 2.7.

However, there is a price to be paid for the saving in acceleration requirements, namely the added complication in the system instrumentation, since now three rays have to be dealt with, $OT$, $OT'$, and $OM$, rather than two. Is the price too high? or is it reasonable for the application? This is a question that designers of LOS guidance systems have to solve.

(b) In another modification of LOS guidance, a virtual guided object M' rather than M is required to be on the LOS to $T$, where $M'$ is a point at a distance $L$ in front of $M$, in the direction of its velocity. This seems to be the geometrical rule according to which certain underwater weapons are guided. "Selection of this distance [i.e., $L$] in tactical applications is a function of many parameters" [13].

## 2.4  Guidance Laws

In Chapter 1 we have stressed that a guidance law is the feedback algorithm by which the desired geometrical rule is implemented. In the case of LOS guidance, the immediate meaning of the distinction between the geometrical rule and the guidance law is that in the former we have, by definition, but *one* LOS, whereas in the latter we have *two*, $OT$ and $OM$, and we wish to make them coincide.

The obvious choice for the 'error' $e$ in the guidance loop is the deviation of $OM$ from $OT$. In the planar case, $e$ would be chosen to be the difference $\phi = \lambda_T - \lambda_M$ (Fig. 2.16(a)) or, possibly, an odd function of $\phi$. For reasons that will be explained in Sec. 2.4.1(c), $\phi$ is often weighted by the range $r_M$, i.e. $e = r_M\phi$. Preferably we would say that the error is the distance of M from the LOS: this distance is approximated by the product $r_M\phi$.

In the general, nonplanar case, the 'error' can be expressed as an angle in a certain plane. For the reasons alluded to in the last paragraph, in practice the error is usually chosen to be the *cross range* $\mathbf{r}_{M_\perp}$, which is the component of $\mathbf{r}_M$ across

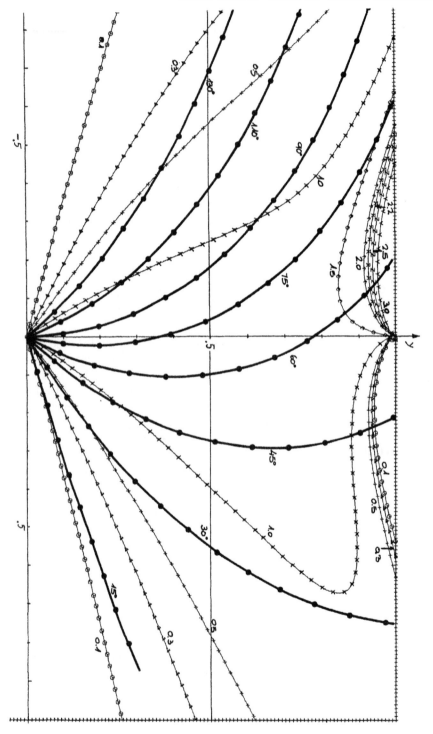

Figure 2.15: Kh plane for modified LOS guidance, $\varepsilon = 0.3$

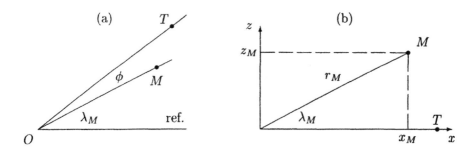

Figure 2.16: Geometry of implementing LOS guidance

$\mathbf{r}_T$. Since $\mathbf{r}_{M_\perp} = \mathbf{1}_{r_T} \times (\mathbf{r}_M \times \mathbf{1}_{r_T})$, it follows that that $|\mathbf{r}_{M_\perp}|$ equals $r_M$ multiplied by the sine of the angle between $\mathbf{r}_M$ and $\mathbf{r}_T$.

This section will deal with planar situations. Two approaches will be shown, namely time-domain and classical control, and references will be given to a third one, modern control.

## 2.4.1 Time-Domain Approach

**(a)** Suppose we are dealing with a stationary T (or, for that matter, a directly-approaching or directly-receding T). Then $\lambda_T$ is constant and may conveniently chosen to be zero, so that the LOS $OT$ coincides with the reference line $Ox$. M has the coordinates $(x_M, y_M, z_M)$ with $y_M \equiv 0$ due to the planar engagement assumption (Fig. 2.16(b)). By convention, the 'error' is denoted by $\phi$ and is defined by the equation

$$\phi \overset{\Delta}{=} \lambda_T - \lambda_M , \qquad (2.37)$$

which, due to the definition $\lambda_T \overset{\Delta}{=} 0$, reduces to $\phi = -\lambda_M$. We assume that M's dynamics is lag-free, or zero-order, such that $a_M \equiv a_{M_c}$.

The most obvious guidance law, of course, is to command M to maneuver in direct proportion with $\phi$, i.e.,

$$a_{M_c}(t) = A\phi(t) , \qquad (2.38)$$

where $A$ is a constant. Assuming that $|z/r| \ll 1$, where $r = r_M$ and $z = z_M$ for the sake of brevity, the approximation

$$\phi = -\lambda_M \approx -\frac{z}{r} \qquad (2.39)$$

is valid, as well as the assumption that $a_M = \ddot{z}$ is parallel to the $z$-axis.

The guidance law (2.38) now leads to the differential equation

$$\ddot{z}(t) = -A\frac{z(t)}{r(t)} \; . \tag{2.40}$$

If we further assume that M's velocity $v$ is constant, such that

$$r(t) = vt \; , \tag{2.41}$$

and define

$$k \triangleq \frac{A}{v} \; , \tag{2.42}$$

we obtain the differential equation

$$\ddot{z}(t)t + kz(t) = 0 \; . \tag{2.43}$$

Its solution is stated in terms of Bessel function of the first kind and order 1, $J_1(.)$, as follows [14].

$$z(t) \propto \sqrt{4kt}J_1(\sqrt{4kt})$$

("$\propto$" denotes "proportional to"). Using (2.39) and (2.41), we easily modify this result to an expression for $\phi(t)$:

$$\phi(t) \propto \frac{1}{\sqrt{4kt}}J_1(\sqrt{4kt}) \; .$$

For large values of its argument, $J_1(\xi)$ behaves approximately like the function $\frac{1}{\sqrt{\xi}}\cos(\xi + ...)$ [ibid.]. Hence, for large values of $\sqrt{4kt}$, $\phi(t)$ and $z(t)$ are, approximately,

$$\phi(t) \propto t^{-\frac{3}{4}} \cos(\sqrt{4kt} + ...) \tag{2.44}$$

and

$$z(t) \propto t^{\frac{1}{4}} \cos(\sqrt{4kt} + ...) \; . \tag{2.45}$$

Thus, while $\phi$ oscillates with decreasing amplitudes, $z$ diverges! A human observer watching M from $O$ along the LOS would report that M converged to the LOS, since the human eye responds to angles. This, alas, would be only partly true: If M was a low-flying anti-tank missile, then sooner or later it would hit the ground.

(b) It will not come as a surprise to readers acquainted with control theory that the situation may be remedied by adding a rate term, or derivative control, such that a new guidance law is obtained, as follows:

$$a_{M_c}(t) = A(1 + T_d\frac{d}{dt})\phi(t) \; , \tag{2.46}$$

$T_d$ being a constant. The differential equation analogous to (2.43) is

$$t\ddot{\phi}(t) + (2 + kT_d)\dot{\phi}(t) + k\phi(t) = 0 \,, \tag{2.47}$$

whose solution in terms of $z(t) = -vt\phi(t)$ is

$$z(t) \propto (4kt)^{\frac{1-kT_d}{2}} J_{-1-kT_d}(\sqrt{4kt}) \tag{2.48}$$

(see [14] for details). An approximation for (2.48) for large values of the argument $\sqrt{4kt}$ is

$$z(t) \propto t^{\frac{1-2kT_d}{4}} \cos(\sqrt{4kt} + ...) \,, \tag{2.49}$$

according to which a necessary and sufficient condition for convergence is

$$T_d > \frac{1}{2k} \,. \tag{2.50}$$

(c) Another remedy for the stability problem that evolves from the 'obvious guidance law' (2.38) requires estimating $z$ from $\phi$ and $r$, the latter being either measured or estimated. (since $r(t) = \int v(t) \, dt$ and $v(t)$ is generally roughly known, estimating $r$ is straightforward). After derivative control has been introduced (cf. (2.46)), the guidance law is

$$a_{M_c}(t) = -A'(1 + T_d'\frac{d}{dt})z(t) \,, \tag{2.51}$$

where $A'$ and $T_d'$ are constants. This guidance law directly leads to the very well known second order differential equation

$$\ddot{z}(t) + A'T_d' \dot{z}(t) + A'z(t) = 0 \,. \tag{2.52}$$

$A'$ and $T_d'$ are chosen such that its solution for $z(t)$ is

$$z(t) \propto \epsilon^{-\zeta\omega_n t} \cos(\omega_n t\sqrt{1 - \zeta^2} + ...) \,,$$

$$\omega_n = \sqrt{A'} \,, \quad \zeta = \frac{T_d'}{2}\sqrt{A'} < 1 \,,$$

and $\omega_n$, the natural frequency, and $\zeta$, the damping coefficient have satisfactory values.

*Note.* In beam-riding LOS guidance systems, $z(t)$ is indeed measured *directly* (by M), and neither the error angle $\phi$ nor the range $r$ is directly involved in the guidance process.

(d) In conclusion of this subsection, we observe that analytic time-domain approach to the study of basic LOS guidance laws is attractive from the didactic point of view, being direct and using well known mathematical methods. However, even under the assumption of lag-free dynamics it has proven to be somewhat complex. If more complicated dynamics is assumed, this approach becomes quite unpractical.

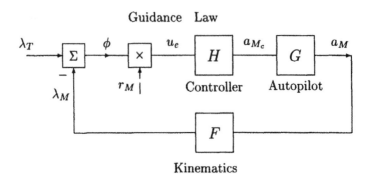

Figure 2.17: LOS guidance loop

## 2.4.2   Classical-Control Approach

**(a)** In the manner of control engineers, we start the discussion with a block diagram, Fig. 2.17, which shows the signals that exist in the guidance loop. The symbols that appear in the block diagram have already been defined, with the exception of $u_e$, the input to the controller, and $G$ and $H$, transfer functions of the controller and the dynamics, or autopilot, respectively. (In all of Sec. 2.4.1 we assumed $G = 1$ (lag-free dynamics); In Sec. 2.4.1(a) and (b) we had $u_e = \phi$; in Sec. 2.4.1(b) and (c), the controller transfer function was $H(s) = A(1 + sT_d)$; and in Sec. 2.4.1(c), $u_e$ was $r_M\phi$).

Let us now choose

$$u_e = \phi$$

and examine the block labelled "kinematics" in Fig. 2.17. We recall (2.6) which gave $a_M$ in terms of $\dot{\lambda}$ and $\ddot{\lambda}$, as follows:

$$a_M = 2v_M(\dot{\lambda} + \frac{r_M(\ddot{\lambda} - \dot{v}_M\dot{\lambda}/v_M)}{2\dot{r}_M}) \ . \tag{2.53}$$

Assuming for a moment that $r_M/\dot{r}_M$ and $\dot{v}_M/v_M$ are constant, (2.53) can be expressed as the transfer function

$$\frac{\lambda(s)}{a_M(s)} \triangleq F(s) = \frac{1}{2v_M s(1 - \frac{r_M\dot{v}_M}{2\dot{r}_M v_M} + s\frac{r_M}{2\dot{r}_M})} \ , \tag{2.54}$$

which, with some reservations, is inserted into the block diagram, Fig. 2.17, as representing the kinematics. The reservations of course are due to the fact that the coefficients in the parantheses of (2.54) are not really constant. This F(s) can therefore be used for studies of the loop in which *specific points* in time, or flight conditions, are examined, i.e., for *frozen range* studies, as they are sometimes called.

For each point, the appropriate values of $r_M$, $\dot{r}_M$, $v_M$, and $\dot{v}_M$ should be substituted in $F(s)$.

Some simplification is obtained by assuming that $v_M$ is constant, as we have already done in deriving (2.7) from (2.6), and that $\dot{r}_M = v_M$, which is permissible if the lead angle $\delta$ is small such that $\cos \delta \approx 1$. We obtain a simplified $F(s)$, as follows:

$$F(s) = \frac{1}{2v_M s(1 + sT_e)} \, , \tag{2.55}$$

where $T_e \triangleq r_M/2v_M$ is a time constant (which in fact is not constant, as noted above). $T_e$ *increases with range* (or time); in engagements where the trajectory is approximately a straight line, e.g., in most anti-tank missile systems, $T_e$ roughly equals half the time elapsed from launch.

Experience has shown that, for the purpose of frequency-domain analyses, the inequality $T_e|s| \gg 1$ holds for all but the lowest frequencies [15], so that $F(s)$ then reduces to the variable-gain double integrator

$$F(s) = \frac{1}{r_M s^2} \, . \tag{2.56}$$

The classical approach to the design of a feedback system as described in Fig. 2.17 and where $F$ is given by (2.56) would be to introduce derivative control; for example, by making $H(s)$ have the form

$$H(s) = A\frac{1 + sT_1}{1 + sT_2}, \quad T_1 > T_2 \, , \tag{2.57}$$

with the gain $A$ possibly increasing, to compensate for the varying $r_M$ in the denominator of $F(s)$. Control engineers call such a transfer function a *lead network*; clearly this 'network' is another, more realistic way of obtaining derivative control than (2.46) or (2.51).

The dynamics of the autopilot has still not been considered. 'Autopilot' in Fig. 2.17, as well as in the rest of this book, includes all the devices, electronics, gyros and accelerometers, rudder servos, and so forth, and the aerodynamics that take part in translating the command $a_{M_c}$ to actual acceleration $a_M$. When the dynamics *is* considered, such that $G(s)$ is no more identically equal to 1, more complicated networks will probably be needed than $H(s)$ of (2.57). Finally, it is hardly necessary to say that, having derived a suitable $H$, the designer always carries out extensive simulations in order to check the robustness of the design against various uncertainties in the model.

(b) In presenting classical approach so far we have tacitly assumed $\lambda_T = const$ (or zero), as indeed we have done when studying the time-domain approach in Sec. 2.4.1. Such a simplification may be in order when slow-moving or faraway targets

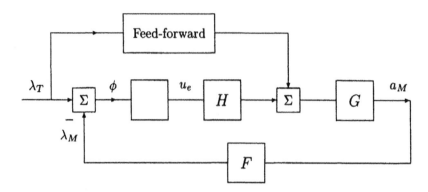

Figure 2.18: LOS guidance loop with feed-forward

are involved, but this is not the case in many other systems, where target motion is fast.

By 'fast' we mean high values of $\dot{\lambda}_T$ and $\ddot{\lambda}_T$ and, more generally, 'high bandwidth' $\lambda_T$ input. When such inputs are considered, important changes must be made to the straightforward elementary design presented above. The changes would necessarily make the loop more 'stiff', i.e., with higher gain, and faster, i.e., possessing higher bandwidth. Such systems tend to be sensitive to noise, always present in real guidance systems, and sometimes not to have sufficient stability margins.

In order to alleviate such difficulties, *feed-forward* compensation is often employed (Fig. 2.18). When this is done, the guidance loop can be low-gain, low-bandwidth, while the system as a whole responds well to fast inputs. This kind of compensation is, in fact, an open-loop addition to the closed guidance-loop. Ideally it should counteract the LOS motions: it has indeed been called the "inverse transfer function to the kinematics" [15]. The role of the feed-forward element is to calculate the maneuvers required in order that M can stay on the LOS in spite of the changes in $\lambda_T$. The calculation is based on the measured $\lambda_T$ and its derivatives.

With or without the feed-forward, the guidance laws employed are, in principle, the same as in Sec. 2.4.1.

## 2.4.3   Optimal-Control Approach

Optimal-control theory dates from about 1960. It began to be used for the study of guidance a few years later, especially in the context of finding an alternative to classical proportional navigation for two-point guidance systems (see Chapter 8). Applying this theory to LOS guidance seems to have started a few years later [16, 17]; in particular, one should mention a series of articles by authors affiliated with

Queen's University of Belfast and with Short Brothers Ltd., also of Belfast [18-21], in which CLOS of small bank-to-turn (BTT)-controlled missiles is dealt with. (On BTT control vs. skid-to-turn (STT) control, see Appendix C). The reader is referred to Chapter 8 for some details on modern control theory applied to guidance problems, and to the references cited above for specific applicaions to LOS guidance.

## 2.5 Mechanization of LOS Guidance

### 2.5.1 CLOS vs. Beam-Riding Guidance

**(a)** No matter what guidance law is employed for a *CLOS guidance system*, the following functions have to be mechanized (see Fig. 2.19(a)).

(1) Tracking T and measuring $\lambda_T$ (and often, $\dot{\lambda}_T$ as well).

(2) Tracking M and measuring $\lambda_M$ and, where relevant, $r_M$.

(3) Calculating the 'error' $\phi = \lambda_T - \lambda_M$.

(4) Applying the guidance law by computing $a_{M_c}$ based on either $\phi$ or $r_M\phi$ and, where feed-forward is used, on $\lambda_T$ and its derivatives as well.

(5) Transmitting $a_{M_c}$ to M.

(6) (At M) receiving the commands $a_{M_c}$.

(7) (At M) Translating them to steering commands.

The list of functions has been formulated for a planar engagement. In some applications this would be quite realistic, for example if M is a boat, or if it is a missile flying over the sea at a constant altitude and being guided in azimuth only. In other cases, each of the variables $\lambda_T$, $\lambda_M$, $\phi$, $z$, and $a_{M_c}$ consists in fact of two components, say in the $y$ and $z$ directions, where $x$ is the forward direction. There must exist means, e.g., control of M's roll angle, for ensuring that the $y$ and $z$ axes at M agree with the $y$ and $z$ axes at the control point where the $a_{M_c}$ commands are determined.

**(b)** *Beam Riding* does not of course require transmitting commands to M. In beam riding (BR) systems, M 'rides' a beam, or a ray, directed at T from O. The ray used to be a radio beam; in recent years, the ray is often an infra-red (IR) or a visual-wavelength beam generated by lasers. An obvious disadvantage of IR and visual-wavelength techniques is that they cannot be used in bad weather.

In BR systems, there are only the following functions (see Fig. 2.19(b)).

(1) Tracking T and generating a beam directed at it.

(2) (At M) measuring the distance $z$ of $M$ from the center of the beam.

(3) (At M) Applying the guidance law by computing $a_{M_c}$, based on $z$.

(4) (At M) Translating $a_{M_c}$ to steering commands.

**(c)** Comparing CLOS to BR mechanizations of LOS guidance shows that in the latter, less complicated *system control* instrumentation is required, at the cost of somewhat more sophistication at M. It depends on the specific application which

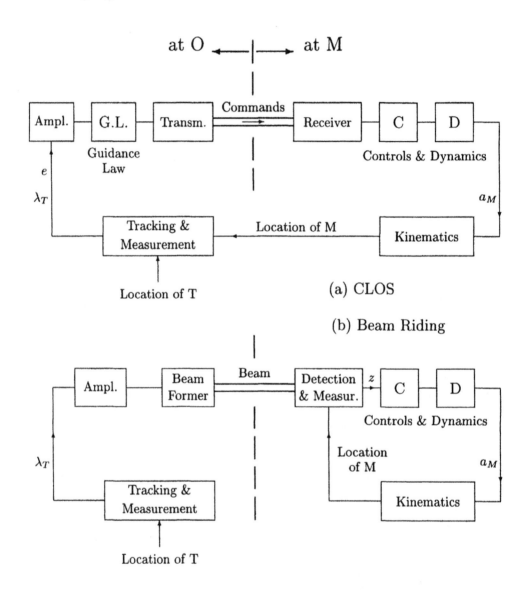

Figure 2.19: Structure of main types of LOS guidance systems

of the two is to be preferred; however, both have been utilized since World War II, for a variety of applications: air defense, anti-tank, and anti-ship [7, 23-25].

## 2.5.2 On Tracking and Seekers

**(a)** In the first generation of CLOS guided missiles (and guided boats, etc.), tracking was manual. An operator had to keep the image of T in the *field of view (FOV)* of a certain device, which could be simple binoculars, a radar scope, or a television (TV) screen. The same operator, or another one, would try to 'cover' the image of T by that of M, both seen simultaneously in the FOV, thereby implementing the geometrical rule. This was done by a joystick, the output of which was processed at a control unit and then transmitted as commands to M.

In BR systems, the task of the operator was much simpler: all he had to do was to track T, thus keeping the beam pointed at it.

In both CLOS and BR systems there exists the delicate preguidance stage, in which the angle difference $\lambda_T - \lambda_M$ must be made sufficiently small in CLOS systems, or, in BR systems, M must be brought sufficiently close to the beam $OT$. Only when this initial stage has successfully been completed can guidance proper begin.

**(b)** In the second generation of CLOS guided weapons, only tracking T is manual; all the other functions, including tracking M, are performed automatically. M would generally be tracked by radar in large systems and by IR devices in small ones. In the latter case, the IR 'signature' of M may have to be enhanced.

In the third generation, the whole process is automatic.

**(c)** Automatic tracking is done by mounting a *seeker* on a *platform* that has sufficient mechanical freedom for the application: in most cases, two degrees of freedom, e.g., azimuth and elevation, are sufficient. An example dealing with the kinematics of a tracking platform has been shown in Sec. 2.3.4. The tracker, i.e., platform and seeker, is usually located near the start point of M; its geometrical center is indeed the reference point $O$ of the guidance process.

In two-point guidance systems, with which we shall deal in Chapters 3-8, the seeker, also called *homing head*, is of course located at M. It is usually mounted on a *gimbal* which is the analog of the platform mentioned above. More on gimbal-mounted seekers will be said in Sec. 6.4.3.

**(d)** A seeker has to detect an object, say T, 'lock' on it, and track it whithin its FOV. It consists of a *sensor*, an *error-measuring device* (EMD), and auxilliary equipment (Fig. 2.20(a)).

The sensor is a device which is sensitive to some energy radiated by T. This device could be

⋆ a TV camera tube (these days, rather a TV charge-coupled device, CCD);

⋆ an IR detector, of which there are many kinds, depending on wavelength and

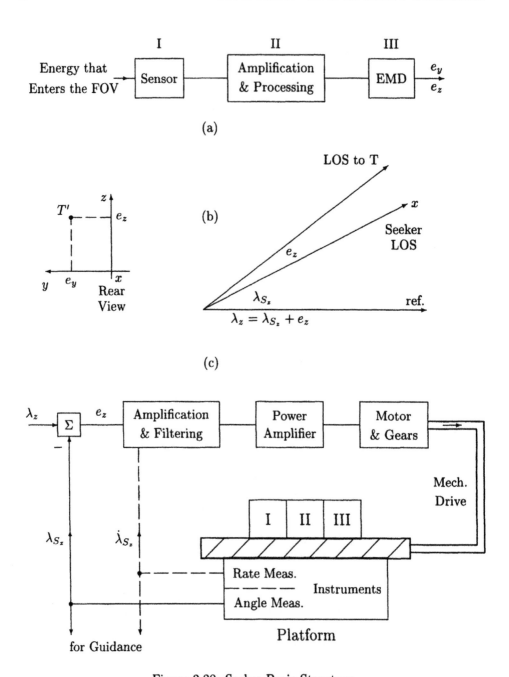

Figure 2.20: Seeker Basic Structure

other parameters;

★ a radio-frequency (RF) receiver for passive and semiactive homing;

★ a radar for active homing;

★ an acoustic detector for underwater weapons;

or, finally,

★ a combination of two of the above.

The input to the EMD is a signal from the sensor which has been processed, e.g, filtered or amplified. The output is the *tracking error e*, i.e., the coordinates $e_y, e_z$ of $T'$, $T'$ being the image of T in the seeker FOC.[2] The $x$-axis of this FOC passes through the center of the FOV in the forward direction (Fig. 2.20(b)). Thus, $e_y$ and $e_z$ are the tracking errors in the $y$ and $z$ channels of the tracker, respectively. Although they are angles they are represented in the FOC diagram in the figure as if they were displacements.

The said tracking errors are used for closing the *tracking loops* of the platform, whose structure depends on the application. A typical tracking loop for a ground-tracker would have the structure shown in Fig. 2.20(c) for the $z$-channel only. As shown in the diagram, a rate-measuring device, e.g., a tachometer is often mounted on the platform; its output, $\dot{\lambda}_S$, may be used both for improving the performance of the tracking loop and for the purposes of the guidance proper (see Sec. 2.4.2(b)).

If the tracker is mounted on a nonstationary base, e.g., a ship, or if the seeker is gimbal-mounted on a moving guided object, e.g., a missile, gyro stabilization is necessary. This kind of application is dealt with in Sec. 6.4.3.

For brief overall descriptions of seekers and target-detection problems the reader is referred to Chapter 2 of Eichblatt and to Chapters 1 and 8 of Garnell [24]. See also the further-reading list for this chapter.

### 2.5.3  Mechanization in Practice

There are several important problems involved with the mechanization of LOS guidance systems that have not been mentioned in this section, as we consider their treatment to be beyond the scope of this text. The main ones are as follows.

★ Tracking T and M is always done in the presence of *noise*, which depends on the type of 'hardware' used, the respective ranges, the atmosphere, etc.

★ The *dynamics* of M's control, i.e., the nature of the transfer function $G$, Figs. 2.17 and 2.18, have been ignored in Sec. 2.4 except for a very brief allusion in Sec. 2.4.2(a). Actually $G$ must be assumed to be a first- or higher-order transfer function, and the effects of the dynamics on the performance of the system may be quite detrimental, particularly so if the dynamics is slow relative to the guidance process.

---

[2]Indeed, a Russian term for a seeker is 'koordinator'.

⋆ Throughout this chapter, we have assumed M's control to be linear. In practice, of course, one has ever-present *nonlinearities* on top of the the dynamics, e.g., asymmetries at the seeker, saturations of fin servos, limits on lateral accelerations. The performance of the real system is in general inferior to that of the analogous linear one.

For a treatment of these and similar problems the reader is referred to Chapter 6, which deals with the mechnization of Proportional Navigation guidance systems, and to the further-reading list for this chapter.

## 2.6   References

[1] von Siemens, Werner, *Inventor and Entrepreneur, Recollections of Werner von Siemens*, 2nd edition, London, Lund Humphries, 1966.

[2] Weyl, A. R., "On the History of Guided-Weapon Development", *Zeitschrift für Flugwissenschaften*, Vol. 5, 1957, pp. 129-138.

[3] Weyl, A. R., *Engins téléguidés*, Dunod, 1952, a translation of *Guided Missiles*, London, Temple Press, 1949.

[4] Benecke, Th. and A. W. Quick (eds.), *History of German Guided Missile Development*, AGARDograph AG-20, AGARD/NATO (Advisory Group for Aerospace Research and Development, North Atlantic Treaty Organisation) First Guided Missile Seminar, Munich, April 1956.

[5] Müller, Ferdinand, *Leitfaden der Fernlenkung*, Garmisch-Partenkirschen, Deutsche RADAR, 1955.

[6] Benecke, Theodor et al., *Die Deutsche Luftfahrt—Flugkörper und Lenkraketen*, Koblenz, Bernard und Graefe, 1987.

[7] Trenkle, Fritz, *Die Deutschen Funklenkverfahren bis 1945*, Heidelberg, Alfred Hüthig, 1987.

[8] Wilder, C. E., "A Discussion of a Differential Equation", *Am. Math. Monthly*, Vol. 38, January 1931, pp. 17-25.

[9] Dantscher, Josef, "Guided Missiles Radio Remote Control", in [4], pp. 109-134.

[10] Schultz, Robert L., *Simplified High Accuracy Guidance System*, US Patent No. 4,146,196, March 1979.

[11] Lebedev, A. A. and L. S. Chernobrovkii, *Dinamika Pol'ota Bespilotnykh Letatel'nykh Apparatov*, Moskva, Oborongiz, 1962.

[12] Weiss, M. and U. Reychav, "Kinematics of a Leading-Beam Rider", *RAFAEL Report* No. 75/27, Sept. 1975.

[13] Bessachini, A. F. and R. F. Pinkos, "A Constrained Fuzzy Controller for Beam Rider Guidance", Newport, Rhode Island, *Naval Undersea Warfare Center Division*, May 1993, AD-A269114.

[14] Jahnke, E., F. Emde, and F. Lösch, *Tables of Higher Functions*, 6th ed., Mc-GrawHill, 1960.

[15] East, D. J., "Some Aspects of Guidance Loop Design for SAM Missiles", *AGARD Lecture Series* No. 135, May 1984.

[16] Kain, J. E. and D. J. Yost, "Command to Line-of-Sight Guidance: A Stochastic Optimal Control Problem", *J. Spacecraft*, Vol. 14, No. 7, 1977, pp. 438-444.

[17] Durieux, J. L., "Terminal Control for Command-to-Line-of-Sight Guided Missile", *AGARD Lecture Series* No. 135, May 1984.

[18] Roddy, D. J., G. W. Irwin, and H. Wilson, "Optimal controllers for bank-to-turn CLOS guidance", *IEE Proc.*, Part D, Vol. 131, No. 4, 1984, pp. 109-116.

[19] Roddy, D. J., G. W. Irwin, and R. J. Fleming, "Linear Quadratic Approach to Optimal Control and Estimation in Bank-to-Turn CLOS Guidance", *IEE Conference Publication*, No. 252, Vol. 2, 1985, pp. 458-463.

[20] Fleming, R. J. and G. W. Irwin, "Filter controllers for bank-to-turn CLOS guidance", *IEE Proc.*, Part D, Vol. 134, No. 1, 1987, pp. 17-25.

[21] Irwin, G. W. and R. J. Fleming, "Analysis of coloured filter controllers for bank-to-turn CLOS guidance", *IEE Proc.*, Part D, Vol. 135, No. 6, 1988, pp. 486-492.

[22] Hexner, G. and M. Ronen, "A combined classical-modern approach to the development of a CLOS guidance law", a lecture given at a seminar organized by the *Israel Institution for Automatic Control*, April 1991.

[23] Clemow, J., *Missile Guidance*, London, Temple Press, 1962.

[24] Garnell, P., *Guided Weapon Control Systems*, 2nd ed., Pergamon Press, 1980, Sec. 7.2.

[25] Ogorkiewicz, R. M., *Technology of Tanks*, Coulsdon, Surrey, U.K., Jane's, 1991, Chapter 9.

## FURTHER READING

Hovanessian, S. A., *Radar System Design and Analysis*, Dedham, Mass., Artech House, 1984.

Wolfe, W.L. and G.J. Zissis (ed.), *The Infrared Handbook*, rev. ed., Washington, D.C., Office of Naval Research, Department of the Navy, 1985.

James, D.A., *Radar Homing Guidance for Tactical Missiles*, Macmillan, 1986.

Hovanessian, S. A., *Introduction to Sensor Systems*, Norwood, MA, Artech House, 1988.

Eichblatt Jr., E. J. (ed.), *Test and Evaluation of the Tactical Missile*, Washington D.C., AIAA, 1989.

Lin, C. F., *Modern Navigation, Guidance, and Control Processing*, Prentice-Hall, 1991.

Seyrafi, Khalil and S. A. Hovanessian, *Introduction to Electro-Optical Imaging and Tracking Systems*, Norwood, MA, Artech House, 1993.

# Chapter 3

# Pure Pursuit

## 3.1 Background and Definitions

When we leave LOS guidance and start discussion of pure pursuit we are changing of course from three-point to two-point guidance. We now have but one range $\mathbf{r}$ and one line of sight $MT$, along which $\mathbf{r}$ is directed.

The geometrical rule of *pure pursuit* (PP) is simply 'Let the pursuer M direct itself at the traget T'. More precisely, 'Let the velocity vector $\mathbf{v}_M$ coincide with $\mathbf{r}$', i.e., with the LOS (Fig. 3.1). This explains the name *scopodrome* sometimes used for PP (skopien = to observe, dromos = act of running). Other names for this geometrical rule are *hound-hare pursuit* (for reasons explained later in this chapter) and just *pursuit*.

In mathematical terms, PP requires that the vector product $\mathbf{v}_M \times \mathbf{r}$ be zero (but such that $\mathbf{v}_M \bullet \mathbf{r} > 0$, otherwise one would have 'pure escape' rather than pure pursuit.) Note that PP is instantaneously planar by definition, the *engagement plane* being defined by $\mathbf{r}$ and $\mathbf{v}_T$. If T is not maneuvering, then this plane is fixed and the engagement is planar.

It is no wonder that the first generation of two-point guided weapons utilized

Figure 3.1: An illustration for the definition of pure pursuit

this simple rule. In fact, as noted in Sec. 1.2.2, 'homing' was synonymous with 'pure pursuit' until the early 50's, both in the United States and the Soviet Union [1, 2, 4].

## 3.2  Some of the Long History

For two hundred years, PP was a source of delight to mathematicians, especially to amateurs of recreational mathematics. A great number of articles, papers, and notes have been published; an excellent historical review is Clarinval's, published under the auspices of UNESCO [5]. The four articles by Bernhart form an outstanding treatise [6-9]. Pursuit curves seem to be aesthetically pleasant and to have inspired some 'mathematical art' [10, 11]. A few modern references are given by Bruckstein in a very special paper [12].

It all started with a paper presented by the French hydrographer and geometer Pierre Bouguer to the Royal (French) Academy of Sciences in 1732. He formulated the problem in the setting of a pirate boat trying to intercept a merchantman, and presented its solution in an 'absolute' frame of coordinates, i.e., as drawn on a marine chart, say.

The term 'hound-hare' began to be used much later. The origin is no doubt in a note by Dubois-Aymé, a French customs official, published in 1811. He solved the problem again—not knowing about Bouguer's paper—having been prompted by watching the traces left by his dog when chasing him on the beach. Indeed it does seem that many predators catch their prey following the PP geometrical rule.

Applying this rule to practical weapon systems dates from the second world war. The first application resulted from the fact that, while in 'dogfights', aircraft pilots very often executed pure pursuit when chasing their targets [13, 14]. (When attempting to fire, they actually executed *lead pursuit*, since the fighter's gun must be pointed ahead of the target at a certain *lead angle* [15]). Judging by literature published, interest in this kind of application of PP lasted at least until the late 1960's [16, 17].

The second application resulted from the drawbacks of the veteran LOS guidance method. One of the big problems involved with LOS guidance is that accuracy deteriorates with range (see Sec. 2.3.3). Thus, the German Luftwaffe, in spite of initial operational success obtained with the LOS-guided bombs Hs-293 and Fritz-X (see Sec. 2.2), started programs of converting the former to a PP-guided bomb, eventually called Hs-293D. It should be added that the reason for this was probably not solely insufficient accuracy; it also was that CLOS guidance turned out to be dangerous for the aircraft from which the bomb was guided.

In the improved bomb (Hs-293D), a TV camera was installed [19]. Via a radio link, the operator saw the target at *decreasing* ranges, which greatly improved the accuracy, and transmitted commands to the bomb, such that the image of the target would remain at the center of the TV field-of-view. The attacking aircraft could in the meantime get away from the danger zone. This scheme made the Hs-293D a command-guided PP-bomb.

In an American development program, trained pigeons were to be used rather than TV [20]. Antithetically, it is probable that the Japanese *kamikaze* pilots executed PP during at least a part of the trajectory of their *Baka* aircraft. The same applies to the less known German suicide pilots (*Selbstopferpiloten*, literally self-sacrifying pilots). The idea of deploying human-piloted flying bombs was originated in Germany by Hanna Reitsch and Erich Lange in the middle of 1943. The suicide aircraft was a modification of the Fieseler 103, better known as the V-1 flying bomb. Although 175 piloted Fi-103's were produced, only 34 pilots had finished their training by 24 February, 1945. Unlike the Japanese weapon, the German one never became operational [21].

In other weapon development programs, seekers were studied based on radio-frequency (RF) receivers (in the US as well as in Germany [22]) for homing onto radar stations and radio transmitters; infra-red (IR) detectors for homing on heat-radiating objects; and acoustic detectors [23-25]. The guidance for all of these programs would have been according to the PP geometrical rule.

Finally, we note that most of the "smart weapons" (bombs and missiles) which were used extensively in the Viet-Nam and the Gulf wars utilized the PP geometrical rule, homing on laser-illuminated ground targets. More recently, smart bullets based on the same principles have also been proposed [35].

## 3.3  Kinematics

In this section we shall study the kinematics of trajectories produced when M pursues T following the PP geometrical rule. We shall first examine the case of planar engagements, i.e., engagements that take place on a fixed plane.

### 3.3.1  The Planar Case, Nonmaneuvering T

(a) Let our plane be the $z = 0$ plane of a Cartesian frame of coordinates (FOC). Without loss of generality we assume that T moves along the line $x_T(t) = c$, $y_T(t) = v_T t$ (Fig. 3.2). Suppose M starts pursuing T from the origin, i.e., $x_M(0) = y_M(0) = 0$, and that the velocity ratio $K = v_M/v_T$ is constant. Bouguer has shown that M's trajectory $y_M(x_M)$, or $y(x)$ for the sake of brevity, is given by the equation

$$\frac{y}{c} = \frac{K}{K^2 - 1} \left\{ 1 + \frac{1}{2} \left[ (K - 1) \left( 1 - \frac{x}{c} \right)^{\frac{K+1}{K}} - (K + 1) \left( 1 - \frac{x}{c} \right)^{\frac{K-1}{K}} \right] \right\} \qquad (3.1)$$

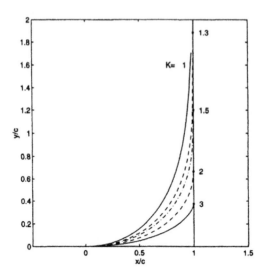

Figure 3.2: Trajectories of pure pursuit

for $K \neq 1$, and by the equation

$$\frac{y}{c} = \frac{1}{4}\left[\left(1 - \frac{x}{c}\right)^2 - \log\left(1 - \frac{x}{c}\right)^2 - 1\right] \tag{3.2}$$

for $K = 1$ [26-28]. (The derivation of (3.1) and (3.2) is somewhat long. Since it is not of much importance to the text that follows, the reader is referred to the said references for a complete treatment.) We refer to these equations as the solution in an *absolute* FOC, since the $(x, y)$ FOC may be regarded as inertial; we shall later use FOC's that are, so to speak, attached to T, and refer to them as *relative*. Trajectories obtained by (3.1) and (3.2) are shown in Fig. 3.2 for five values of $K$.

Several interesting properties result from (3.1) and (3.2), as follows.

(i) All the trajectories tend to T's path $x = c$.

(ii) Intercept is attained if, and only if, $K > 1$. The point of intercept is $x_f = c, y_f = [K/(K^2 - 1)]c$ and the trajectory of M is tangent to that of T at that point. If $v_T$ is constant, the total time of guidance is

$$t_f = \frac{y_f}{v_T} = \frac{K}{K^2 - 1}\frac{c}{v_T}.$$

(iii) If $K = 1$, M tends to a point $c/2$ directly behind $T$.

(iv) If $K < 1$, M still tends to the line $x = c$, but the distance $MT$ increases indefinitely.

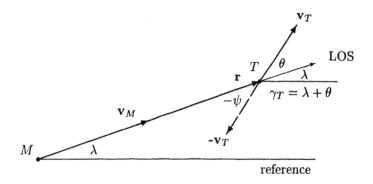

Figure 3.3: Geometry of planar pure pursuit

The same kind of conclusions are obtained for other initial conditions. The trivial case of head-on encounter is an exception on which no time need be wasted.

*Note.* In *all* PP trajectories, the LOS $MT$ is tangent at $M$ to the trajectory of M. This property is an immediate result of the definition of this geometrical rule.

**(b)** We now wish to study the kinematics of PP in a *relative* rather than absolute FOC, i.e., we shall seek the location of M in a FOC attached to T. Thus, a trajectory in the relative FOC shows the situation as seen by an observer located at T. The geometry is depicted in Fig. 3.3.

Since $\dot{\gamma}_T = 0$ by assumption (as T is not maneuvering) and $\lambda + \theta = \gamma_T$ by definition, it follows that

$$\dot{\theta} = -\dot{\lambda} \ . \tag{3.3}$$

By (3.3), (A.13) and (A.14) we have the following equations:

$$\begin{cases} \dot{r} = v_T \cos\theta - v_M \\ \dot{\theta} = -\dot{\lambda} = -\frac{v_T}{r}\sin\theta \ . \end{cases} \tag{3.4}$$

Assuming that K is constant, there results the differential equation

$$\frac{dr}{d\theta} = \frac{r(K - \cos\theta)}{\sin\theta} \ . \tag{3.5}$$

We now define a particular relative FOC which will be referred to as the *target's tail* (TT) FOC. The reference axis in this FOC is the direction of $-\mathbf{v}_T$ (hence 'tail') and the location of M in this FOC is given by the polar coordinates $(r, \psi)$ where, by definition,

$$\psi \overset{\Delta}{=} -\theta \tag{3.6}$$

(see Fig. 3.3). The TT FOC is widely used in guidance literature.

Without loss of generality, $\psi = 90°$ (or $\theta = -90°$) and $r = r_{rbm}$ (for *right beam*) are chosen for the initial conditions; by separation of variables and integration, the solution to (3.5) is found to be

$$r(\theta) = r_{rbm} \frac{\sin^{K-1} \frac{\theta}{2}}{2 \cos^{K+1} \frac{\theta}{2}} = r_{rbm} \frac{\tan^K \frac{\theta}{2}}{\sin \theta} . \tag{3.7}$$

Note that $\theta$ and $\psi$ are interchangeable in (3.5), hence in (3.7) too, such that (3.7) can also be expressed as

$$r(\psi) = r_{rbm} \frac{\tan^K \frac{\psi}{2}}{\sin \psi} .$$

A few interesting properties can be deduced from this equation, as follows.

(i) A trajectory remains on the half-plane where it started (see solid lines in Fig. 3.4).

(ii) Intercept is achieved if, and only if, $K > 1$, such that as $r \to 0, \psi \to 0$ as well: M always ends up in a tail chase.

(iii) For $K = 1$, (3.7) becomes the equation of the parabola $r(\psi) = r_{rbm}/(1 + \cos \psi)$, such that $r \to r_{rbm}/2$ and $\psi \to 0$ as $t \to \infty$.

(iv) For $K < 1$, $r \to \infty$ and $\psi \to 0$ as $t \to \infty$. $r(\psi)$ is not monotone, though; it has a minimum at $\psi = \arccos(K)$ whose value is

$$r_{rbm} \sqrt{[(1 + K)^{1+K}(1 - K)^{1-K}]^{-1}}$$

(see for example the $K = 0.8$ and $K = 0.9$ trajectories in Fig. 3.4).

(c) For $v_T = const$, the following expressions for time (functions of $r$ and $\theta$) are obtained from (3.4).

$$t = \frac{r_0}{v_T} \frac{K + \cos \theta_0 - (\frac{r}{r_0})(K + \cos \theta)}{K^2 - 1} , \quad K \neq 1 , \tag{3.8}$$

$$t = \frac{r_0}{2v_T} \left[ 1 - \frac{1 + \cos \theta_0}{1 + \cos \theta} - (1 + \cos \theta_0) \log \frac{\tan \frac{\theta}{2}}{\tan \frac{\theta_0}{2}} \right] , \quad K = 1 . \tag{3.9}$$

The derivation of (3.8) and (3.9) is deferred to Appendix D.3. In both these equations, $|\theta_0|$ is not necessarily $90°$. However, for $\theta_0 = \pm 90°$, (3.9) reduces to

$$t = \frac{r_0}{2v_T} \left( \frac{\cos \theta}{1 + \cos \theta} - \log | \tan \frac{\theta}{2} | \right) = \frac{r_0}{2v_T} (1 - \frac{r}{r_0} - \log \sqrt{2 \frac{r}{r_0} - 1}) .$$

Based on (3.7), (3.8), and (3.9), *isochrones* for pure pursuit can be derived. (It is recalled that isochrones are lines of equal $t$, i.e., they serve to show how long it

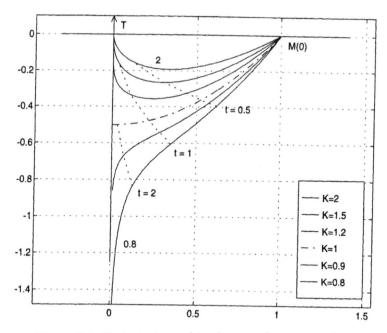

Figure 3.4: Trajectories and isochrones of pure pursuit

has taken M to arrive at a certain point from the start of the pursuit). One has to resort to numerical solutions of these equations in order to draw PP isochrones. The dotted lines in Fig. 3.4 are isochrones for $t = 0.5$, 1, and 2, the unit of time being $r_{rbm}/v_T$.

*Total time of flight* $t_f$ is obtained from (3.8) by substituting $r = 0$:

$$t_f = \frac{r_0}{v_T} \frac{K + \cos\theta_0}{K^2 - 1}, \quad K > 1 \, . \tag{3.10}$$

We can now look for expressions for $t_f$-*isochrones*, i.e., loci of points such that the duration of flight that starts at them is $t_f$. It turns out (see details in Appendix D.4) that the isochrones are the ellipses

$$\left(\frac{x_0 + 1}{K}\right)^2 + \frac{y_0^2}{K^2 - 1} = 1 \, , \tag{3.11}$$

where $x_0$ and $y_0$ are dimensionless TT coordinates, such that

$$x_0 = \frac{r_0 \cos\psi_0}{v_T t_f}, \quad y_0 = \frac{r_0 \sin\psi_0}{v_T t_f}$$

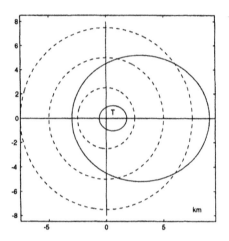

Figure 3.5: Launch zone ellipses

and where the subscript "0" denotes "at $t = 0$". The center of each ellipse is at $x_0 = -1$ and $y_0 = 0$ (i.e., in front of $T$), the major semiaxis and the minor semiaxis being $K$ and $\sqrt{K^2 - 1}$, respectively. $T$ is one of the foci of the ellipse.

An example is taken from missilery. Suppose, as is often the case, that the flight time of a certain air-to-air missile has to be bounded from both below and above, and let the bounds be 3 and 15 *sec*, respectively. For $v_T$ and $v_M$ 200 and $400\,m/sec$, respectively, the missile *launch zone* (also called *capture zone* and *launch envelope*), i.e., the space from which it can be launched such that $t_f$ does not exceed the said bounds, is between two nonconcentric $K = 2$ ellipses, as shown in Fig. 3.5.

*Remark.* Launch zones, a very important concept in missilery, are not determined by bounds on $t_f$ only. They may also be determined by energies available (propulsion fuel, electrical batteries, compressed gas), lateral-acceleration capabilities (which depend on altitude and speed), safety considerations, and other parameters. The inner and outer boundaries of a launch zone are called *minimum range* and *maximum range*, respectively, for obvious reasons.

**(d)** We will now find the *locus of interception*, i.e., the answer to the following question. Suppose T is at the origin of an absolute FOC at $t = 0$, starting motion along a ray that starts at the origin, and suppose M is then at the point $(-r_0, 0)$. Where does M intercept T, given $K > 1$?

Clearly $t_f$ equals $\sqrt{x^2 + y^2}/v_T$, where $x$ and $y$ are the coordinates of the interception point. Equating this expression with (3.10) gives the equation of the locus,

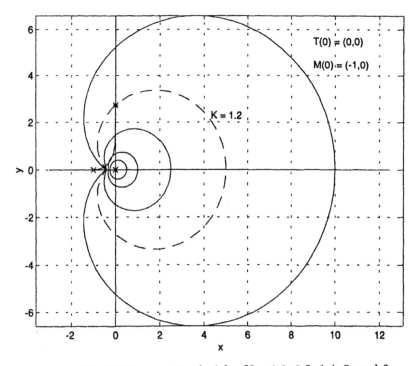

Figure 3.6: PP Interception loci for $K = 1.1$, $1.2$, $1.4$, $2$, and $3$.

which is best described in polar coordinates $(r, \phi)$, as follows:

$$r(\phi) = r_0 \frac{K + \cos \phi}{K^2 - 1} \ . \tag{3.12}$$

This turns out to be the equation of *Pascal's limaçon* (snail).[1]

Interception loci for several values of $K$ are shown for an illustration in Fig. 3.6. Also shown is a pursuit trajectory for $K = 1.2$, $\theta_0 = 90°$. The interception point is indicated by an asterisk ($*$).

(e) Lastly we turn our attention to lateral accelerations $a_M$ required by M during the pursuit.

Since $\gamma_M = \lambda$ in PP and $a_M = v_M \dot{\gamma}_M$, the following expression is easily derived from (3.4) and (3.7) (having substituted $\psi$ for $\theta$):

$$a_M(\psi) = \frac{K v_T^2}{r_{rbm}} \frac{\sin^2 \psi}{\tan^K \frac{\psi}{2}} = \frac{V_T^2}{r_{rbm}} \frac{4K \cos^{K+2} \frac{\psi}{2}}{\sin^{K-2} \frac{\psi}{2}} \ . \tag{3.13}$$

---

[1] This curve was studied by G. Roberval in the middle of the seventeenth century. He named it after Étienne Pascal, Blaise Pascal's father [26].

We deduce the important property that $a_M \to 0$ as $\psi \to 0$ only if $K < 2$; for $K > 2$, $a_M \to \infty$. For $K \le 2$, $a_M(\psi)$ has a maximum at

$$\psi = \arccos \frac{K}{2} \,,$$

whose value is given by the rather cumbersome expression

$$a_{Mmax} = \frac{v_T^2}{r_{rbm}} \frac{K}{4} \sqrt{(2+K)^{2+K}(2-K)^{2-K}} \,.$$

For $K = 1$, the maximum is $1.3 v_T^2/r_{rbm}$, occurring at $\psi = \pm 60°$; for $K = 2$, the maximum is $8 v_T^2/r_{rbm}$, at the intercept ($\psi = 0$).

Another result of (3.4) and the fact that $a_M = v_M \dot{\gamma}_M = v_M \lambda$ in PP is that *isomaneuver*, or equal-$a_M$, lines in the TT relative frame of coordinates are circles given by the equation

$$r(\psi) = \frac{v_M v_T \sin \psi}{a_M} \,.$$

The diameter of each circle equals of course $v_M v_T / a_M$. A family of such circles is shown in Fig. 3.7 for $a_M^* = 0.8, 1, 1.333, 2$, and 4, where $a_M^* \triangleq a_M/[v_M v_T/r_{rbm}]$ is nondimensional maneuver acceleration. The trajectories shown in Fig. 3.4 are shown again here. Since they are drawn only for $\psi$ in the first quadrant, *semi*circles have been drawn rather than full circles. Clearly, maximum value of maneuver is attained by M when its trajectory is tangent to the isomaneuver circle. For $K = 1$, for example, the trajectory is tangent to the $a_M^* = 1.3$ circle; Fig. 3.7 shows the $a_M^* = 1.333$ one.

From the point of view of applications in practice, a very important conclusion can be drawn, namely that whenever $K > 2$, which is very often the case, a PP-guided M will, due to acceleration limitation, 'lose' its target sometime before the expected intercept.

(f) An interesting variation of pursuing a nonmaneuvering target according to the PP geometrical rule is when M keeps the distance $MT$ constant, say $c$, while still following the PP geometrical rule; for example, when a detective (or a bodyguard, if you will) M follows a client T at a discrete constant distance $c$, always 'aiming' at T. For another example, when T tows M, using a constant-length rope.

The relative trajectory, i.e., the trajectory in the TT FOC, is clearly a circle with radius $c$. The trajectory in an absolute FOC depends on the type of motion of T. Suppose, for example, that T moves on a straight line at a constant speed.[2] The trajectory of M in an absolute FOC can be shown to be a *tractrix* curve. For

---

[2] This rope problem was proposed in 1693 by Claude Perrault, and solved — independently and simultaneously — by Leibniz and Huygens a short time later. However, Newton had already formulated the differential equation and solved it in 1676 [26].

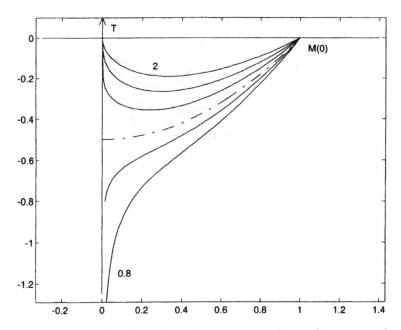

Figure 3.7: Trajectories and isomaneuver lines of pure pursuit

T travelling on the line $x_T(t) = c$, $y_T(t) = v_T t$ and $x_M(0) = y_M(0) = 0$, the parametric equations of the tractrix curve are

$$x_M = c(1 - \sin \theta), \quad y_M = c\left(-\cos \theta + \log \cot \frac{\theta}{2}\right), \quad t = \frac{c}{v_T} \log \cot \frac{\theta}{2}$$

(see Fig. 3.8 ). As $t$ goes from 0 to $\infty$, $\theta$ goes from $90°$ to 0, $x_M \to c$, and $y_M \to y_T - c$.

## 3.3.2 The Planar Case, Maneuvering T

The simplest maneuver, of course, is moving on a circle at a constant speed. The first publication that deals with PP in this context dates from 1742, just ten years after Bouguer's pioneering paper, and appeared on the pages of *Ladies' Diary* [7].

We shall treat the problem in the same way as in Sec. 3.3.1(b). In fact, the reasoning is the same except that here $\dot{\gamma}_T$ is a nonzero constant, say $\omega$, so that

$$\dot{\theta} = \omega - \dot{\lambda}.$$

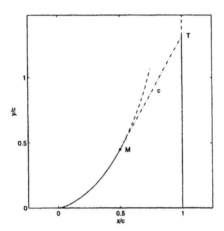

Figure 3.8: Towing by rope, a kind of pure pursuit

Thus, instead of the differential equation (3.5) we now have the differential equation

$$\frac{dr}{d\theta} = \frac{r(\cos\theta - K)}{\frac{r}{c} - \sin\theta} , \qquad (3.14)$$

where $c$ is the radius $v_T/\omega$. "This equation presents unusual difficulties" [28]; however, some results can be derived from it even without solving it, as follows.

(i) $r$ decreases monotonously with $\theta$ if $K \geq 1$.

(ii) Intercept is achieved if, and only if, $K > 1$.

(iii) Intercept is from $\theta = 0$ or $180^o$.

(iv) For $K < 1$, the pursuit approaches a situation called *limit cycle*, where M's trajectory is a circle with radius $Kc$, concentric with T's circle, such that M is at the constant distance $c\sqrt{1 - K^2}$ from T, at the angle $\psi = \arccos K$ off its tail (Fig. 3.9).

An example is borrowed from a 1969 study on 'dogfights' (Fig. 3.10) [16]. Four PP trajectories are shown, for two values of $K$ and for two types of initial conditions. T flies in the manner described above. Interceptor I starts the pursuit either at the center of T's circle, Fig. 3.10(a), or outside that circle, at a distance $c$ from T, Fig. 3.10(b). The initial positions are marked $T(0)$ and $I(0)$. Intercepts occur at $\gamma_{T_f}$, and $\eta$ is the ratio $|a_I/a_T|$. Needless to say, from the point of view of I, the smaller $\eta_{max}$ (and also $\gamma_{T_f}$), the better.

### 3.3.3   Other Interesting Planar Pursuits

We have already noted that many beautiful problems involving PP have been posed—and solved—by amateurs of recreational mathematics. Observing PP in

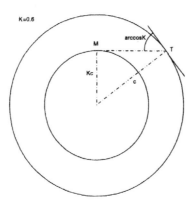

Figure 3.9: Limit cycle produced when $K < 1$

the animal domain (leaving dogs aside) is of particular interest, especially if the creatures are ants. Bruckstein has recently shown that the reason "ants trails look so straight and nice"[3] is that ants may tend to follow one another according to the PP geometrical rule (Fig. 3.11) [12]. Occasinally the ants may get mixed up, and a *cyclic* pursuit results, as shown for example in Fig. 3.12.

A special case of cyclic PP's is where the participants—dogs, bugs or turtles, according to the respective authors' imagination—are arranged at the vertices of regular polygons, all having the same speed and orientation [9]. An illustrative example is shown in Fig. 3.13, where the polygon is a square; it can be shown that the four trajectories are logarithmic spirals, a few words on which will be said in Sec. 3.3.4(e).

### 3.3.4 Deviated Pure Pursuit

(a) It is intuitively clear that M would benefit by pointing its velocity vector at the *future* position of T rather than at the instantaneous one, as done in PP. In other words, it is advantageous for M to have $v_M$ *lead* $r$, where 'lead' means 'in the direction of the future position of T'.[4] The main advantage is in decreasing the lateral accelerations required; a secondary one, reducing $t_f$. If $v_M$ leads $r$ by a *constant* angle, then this is *deviated pure pursuit* (DPP). DPP may also be the

---

[3]The quotation is from Richard P. Feynman's "Surely You're Joking, Mr. Feynman", W. W. Norton, 1985.

[4]Many parents have observed that most children of tender age tend to chase one another following the pure-pursuit rule. (We have said that dogs prefer it, too, hence the term 'hound and hare' pursuit.) Children about three or four years old, however—as well as cats, according to some feline admirers—intuitively use the improved rule described here.

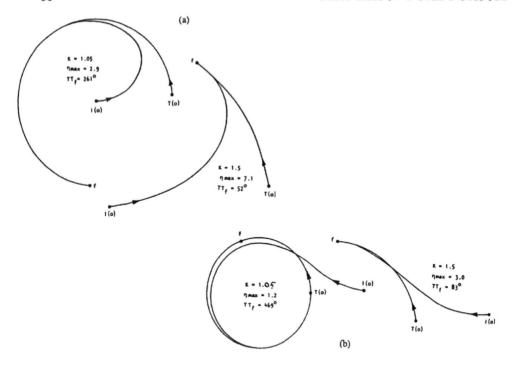

Figure 3.10: Pure pursuit in dogfights (Source: Reference 16)

result of a manufacturing error; in such a case, $\mathbf{v}_M$ may either lead $\mathbf{r}$ or lag behind it.

Note that we have tacitly assumed here that the engagement is *instantaneously planar*, i.e., the vectors $\mathbf{v}_T$, $\mathbf{r}$, and $\mathbf{v}_M$ are coplanar (see Appendix D.5 for the formulation of the DPP geometrical rule in 3-D vector terms).

**(b)** Let $\delta$ be the angle between $\mathbf{r}$ and $\mathbf{v}_M$ (Fig. 3.14). If $\delta$ is in the same sense as $\theta$, then $\mathbf{v}_M$ is said to lead $\mathbf{r}$; in the opposite case, $\mathbf{v}_M$ is said to *lag* behind $\mathbf{r}$. In other words, for $-180° < \theta < 180°$, $sgn(\delta) = sgn(\theta)$ for the first case, $sgn(\delta) = -sgn(\theta)$ for the second one.

By (A.13) and (A.14), the equations for $\dot{r}$ and $\dot{\theta}$ are now

$$
\begin{cases}
\dot{r} = v_T \cos\theta - v_M \cos\delta \\
\dot{\theta} = -\dot{\lambda} = \frac{1}{r}(-v_T \sin\theta + v_M \sin\delta) \ ,
\end{cases}
\tag{3.15}
$$

respectively. Hence the differential equation

$$
\frac{dr}{d\theta} = \frac{r(\cos\theta - K\cos\delta)}{-\sin\theta + K\sin\delta} \ .
\tag{3.16}
$$

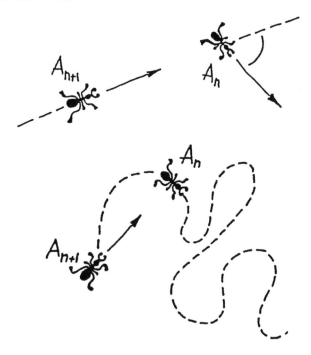

Figure 3.11: Ants utilizing the PP geometrical rule (Source: Reference 12, with kind permission from Springer-Verlag New York, Inc.)

The solution of (3.16) depends on the initial conditions and on the parameters $K$ and $K \sin \delta$. We will examine several interesting cases, assuming $K$ is constant.

(i) For $K > 1$ and $|K \sin \delta| < 1$, which is most often the case, and provided $K \sin \delta \neq \sin \theta_o$,

$$r(\theta) = C \frac{\sin^{\mu-1}\left(\frac{\theta-\beta}{2}\right)}{\cos^{\mu+1}\left(\frac{\theta+\beta}{2}\right)} \tag{3.17}$$

where

$$\beta \overset{\Delta}{=} \arcsin(K \sin \delta) , \tag{3.18}$$

$$\mu \overset{\Delta}{=} K \frac{\cos \delta}{\cos \beta} = \frac{K \cos \delta}{\sqrt{1 - K^2 \sin^2 \delta}} , \tag{3.19}$$

and $C$ depends on the initial conditions. For $\theta_0 = -90^\circ$ and $r_0 = r_{rbm}$, $C = r_{rbm}(1 - \sin \beta)/2$.

M intercepts T from the angle $\psi = -\beta$ (measured from the $-\mathbf{v}_T$ direction; recall that $\psi \overset{\Delta}{=} -\theta$). See, for example, the pursuit curve for $K = 1.3$, $\theta_0 = -90^\circ$, and

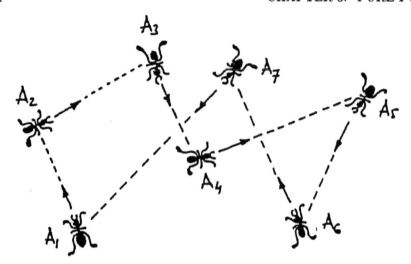

Figure 3.12: An accidental cyclic pure pursuit (Source: Reference 12, with kind permission from Springer-Verlag New York, Inc.)

$\delta = -20^o$ shown in Fig. 3.15. For this case, $\beta = \arcsin(1.3\sin(-20^o)) = -26.4^o$. This of course is a *lead* case not lag one, since $\theta_0$ is negative. In the same figure, a pursuit curve is also shown for pure pursuit, i.e., $\delta = 0$.

(ii) For $K\sin\delta = \sin\theta_o$, the solution is $\theta = 0$ and M's trajectory is a straight line. This is the situation called *collision course conditions*, already mentioned in Sec. 2.3.1(a), to be dealt with in length in Chapter 4. See, for example, the straight line for $K = 1.3$, $\delta = -\arcsin(1/1.3) = -50.3^o$ shown in Fig. 3.15. (Note that all the pursuit curves shown in this figure are for $\theta_o = -90^o$).

(iii) $K > 1$ and $|K\sin\delta| > 1$. In this case, M's trajectory is a spiral which approaches T as $r \to 0$. See for example the curves for $\delta = -51^o$ and $-55^o$ in Fig. 3.15. Note the effect that a relatively small increase of $|\delta|$ from the 'ideal' angle of $50.3^o$ has on the shape of the trajectory.

(iv) $K = 1$. There is no intercept, of course, and the trajectory is easily obtained from (3.16). It is given by the equation

$$r(\theta) = r_0 \frac{1 + \cos(\theta_0 + \delta)}{1 + \cos(\theta + \delta)} \ . \tag{3.20}$$

As $t \to \infty$, $\theta \to \delta$, $\psi \to -\delta$, and $r \to r_0[1+\cos(\theta_0+\delta)]/(1+\cos 2\delta)$. For $\theta_o = -90^o$, M approaches the point

$$\left( \frac{r_{rbm}}{2(1 - \sin\delta)}, -\delta \right)$$

in the $(r, \psi)$ TT (polar) FOC, where it will stay for ever, so to speak. If M and T are

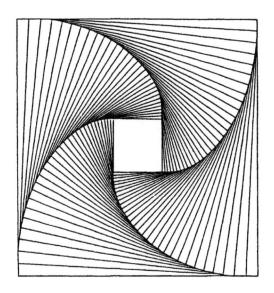

Figure 3.13: A cyclic, symmetrical pure pursuit

aircraft, they are now flying 'in formation'. This is illustrated by the trajectories for $K = 1$ in Fig. 3.15 for $\delta = -20^o$ and 0, respectively. The points marked by "x" signs are where the pursuers M tend when $t \to \infty$.

A comparison between the effects of lead and lag is presented in Fig. 3.16 for $\delta = -20^o$ (lead), 0, and $20^o$ (lag), respectively, and $K = 1.3$. A lag trajectory for $K = 1$ is also shown. The "x" sign has the same meaning as in Fig. 3.15.

(c) The relation between time and $r$ and $\theta$ is obtained from (3.15) in much the same way (3.8) was obtained from (3.4), as follows:

$$t = \frac{r_0}{v_T} \frac{K + \cos(\theta_0 + \delta) - \frac{r}{r_0}[K + \cos(\theta + \delta)]}{(K^2 - 1)\cos\delta} \ , \quad K > 1 \ . \tag{3.21}$$

$t_f$ is obtained from (3.21) by letting $r$ be 0:

$$t_f = \frac{r_0}{v_T} \frac{K + \cos(\theta_0 + \delta)}{(K^2 - 1)\cos\delta} \ , \quad K > 1 \ . \tag{3.22}$$

Based on this equation it can be shown that $t_f$-isochrones are again ellipses (cf. Sec. 3.3.1(c), where $\delta$ was 0 by definition). However, the ellipses are now inclined by the angle $\delta$ with respect to the $-\mathbf{v}_T$ axis, such that the center of each ellipse is at

$$\frac{r_0}{v_T t_f} = \cos\delta \ , \quad \psi_0 = 180^o - \delta$$

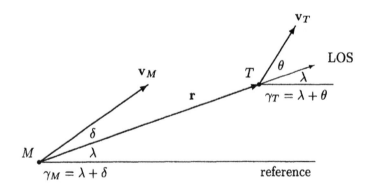

Figure 3.14: Geometry of planar two-point guidance

in the nondimensional TT polar FOC. The major semiaxis and the minor semiaxis are $K \cos \delta$ and $\sqrt{K^2 - 1} \cos \delta$, respectively.

(d) The relation $a_M = v_M \dot{\lambda} = -v_M \dot{\theta}$ is true for DPP just as it is true for PP. Hence, by (3.15), the equation

$$a_M(\theta) = K v_T^2 \frac{\sin \theta - K \sin \delta}{r(\theta)} \ , \tag{3.23}$$

where $r(\theta)$ has already been found.

It can be deduced from the solution that, for $|K \sin \delta| \leq 1$, $a_M(\psi)$ has a maximum at $\psi = \arccos[(K/2) \cos \delta] \equiv \arccos[(\mu/2) \cos \beta]$. Towards intercept (from the angle $\psi = -\beta$, we recall), $a_M \to 0$ as $r \to 0$ if, and only if, $\mu < 2$.

For $\mu = 2$, $a_M \to 4 K v_T^2 / [r_{rbm}(1 - \sin \beta)]$ as $r \to 0$; for $\mu > 2$, $a_M \to \infty$. Thus, in order that M capture T such that the required $a_M$ remains finite as $r$ approaches 0, the inequality

$$1 < K \leq \frac{2}{\sqrt{1 + 3 \sin^2 \delta}} \tag{3.24}$$

must be satisfied.

Generally speaking, for low values of $K$ and $|\delta|$, DPP requires smaller lateral accelerations than ordinary PP if, for $\theta_0 > 0$, $0 < K \sin \delta < \sin \theta_0$. It requires higher ones if $0 > K \sin \delta > -\sin \theta_0$; and conversely for negative $\theta_0$.

(e) The special case of DPP towards a stationary target will now be examined. This special case is not without some practical value in connection with certain weapon systems; also, going back to the realm of animals, it is said that certain night insects fly according to the DPP geometrical rule when approaching light

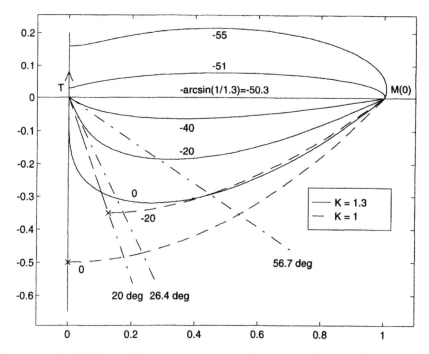

Figure 3.15: DPP and PP trajectories for $K = 1$ and $K = 1.3$, Lead cases

sources (or prey?). From (3.15), the differential equation

$$\frac{dr}{d\theta} = -r \cot \delta \qquad (3.25)$$

is immediately obtained by letting $v_T = 0$. Its solution for $\lambda = -\theta$ is

$$\lambda(r) = \log\left(\frac{r_0}{r}\right)^{\tan \delta} \quad \text{or} \quad r(\lambda) = r_0 \epsilon^{-\lambda \cot \delta}, \qquad (3.26)$$

where $r_0 \triangleq r(0)$. This is the equation of a *logarithmic spiral*.[5] As $r \to 0$, $\lambda$ increases indefinitely, i.e., M approaches T (located at the origin) spiralling an infinite number of times. The shape of the trajectory does not depend on $v_M$; neither does the radius of curvature $\rho$, given by the equation

$$\rho(\lambda) = \frac{r(\lambda)}{\sin \delta}. \qquad (3.27)$$

---

[5]This remarkable curve was found by Descartes in 1638. Jacques Bernouli dubbed it 'spira mirabilis', no doubt due to its marvelous properties. The name 'logarithmic spiral' dates from the beginning of the 18th century.

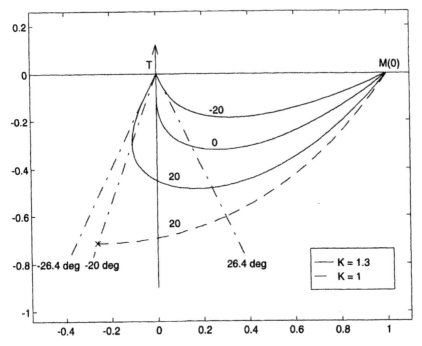

Figure 3.16: Lead vs. Lag DPP for $K = 1$ and $K = 1.3$

$a_M$ is inversely proportional to $\rho$ and directly proportional to $v_M^2$:

$$a_M(\lambda) = \frac{v_M^2}{\rho(\lambda)} = \frac{v_M^2 \sin \delta}{r(\lambda)} .$$

(f) Finally, we turn our attention to the case of *maneuvering targets*. Even for the case where T's maneuver is simply motion on a circle, there is no analytical solution. The equation analogous to (3.14), which was obtained for ordinary PP, is obtained from (3.15) and the geometrical relation $\dot{\theta} = \dot{\gamma}_T - \dot{\lambda} = \omega - \dot{\lambda}$, as follows.

$$\frac{dr}{d\theta} = \frac{r(\cos \theta - K \cos \delta)}{\frac{r}{c} - \sin \theta + K \sin \delta} .$$

Needless to say, it also "presents unusual difficulties". An example that illustrates advantages of DPP is presented in Fig. 3.17 for $K = 1.3$ and $\delta = 20°$. T moves on a circle with radius 1 centered at $(0, 1)$ and M starts its chase either from within the circle or from without it. Two trajectories for ordinary PP (i.e., $\delta = 0$) are also shown for the sake of comparison. "x" and "*" signs indicate points of capture for PP and DPP, respectively.

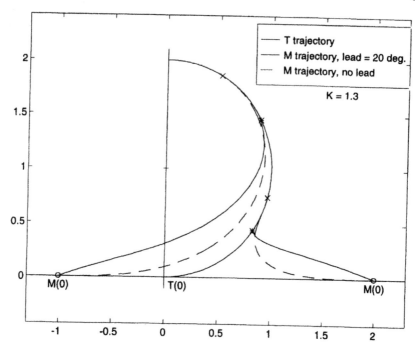

Figure 3.17: DPP vs PP, maneuvering T

## 3.3.5 Examples

### Example 1 - A Homing Boat

Suppose (stationary) T is on the sea shore, located at the origin of the $(x, y)$ plane such that the $y$ axis is along the (straight) coastline (Fig. 3.18). At $t = 0$, M, say a boat, is at the point $(c, 0)$, starting a DPP towards T. The pursuit ends when M hits the shore, i.e., when $x_M = 0$, so that the miss distance is $m = y_M(t_f)$. If there is no lower bound on $\rho$, then $m$ is easily obtained from (3.26) by substituting $\pi/2$ for $\lambda$:

$$m = c\epsilon^{-\frac{\pi}{2}\cot\delta} . \qquad (3.28)$$

If there is such a bound, say $\rho_{min}$, then M may 'saturate' before arriving at the finish line $x = 0$; this happens, of course, if

$$\rho_{min} > \rho\left(\frac{\pi}{2}\right) = \frac{c\epsilon^{-\frac{\pi}{2}\cot\delta}}{\sin\delta} .$$

M would in this case go on maneuvering at it's minimal curvature $\rho_{min}$, and the miss finally produced would exceed the value indicated by (3.28).

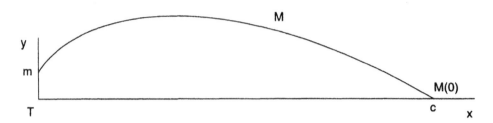

Figure 3.18: Deviated PP with a stationary target, $\delta = 30°$, Example 1 in Sec. 3.3.5

### Example 2 - Landing in Presence of Wind

A controller for vehicles that land in gliding flight uses a rule which, in the horizontal $x, y$ plane, is equivalent to DPP. By this rule, the horizontal component of the *airspeed*[6] $\mathbf{v}_M$ continuously points along the current horizontal radial range $\mathbf{r}$ at a fixed angular offset $\delta$ [30]. Let $u_M$ denote the said horizontal component, and $\mathbf{v}_W$, the wind velocity vector (we assume that the wind is horizontal). Then, the *groundspeed*[5] is given by the vector sum of $\mathbf{u}_M$ and $\mathbf{v}_W$; we will resolve the groundspeed into its radial component $\dot{r}$ and tangential component $r\dot{\theta}$. Assuming that $u_M$ and $v_W$ are constant and that $\mathbf{v}_W$ is directed along the $y$ axis in the negative sense (see Fig. 3.19), it is straightforward to obtain the equations

$$\begin{cases} \dot{r} = -u_M \cos\delta - v_W \sin\theta \\ r\dot{\theta} = -u_M \sin\delta - v_W \cos\theta \, , \end{cases} \tag{3.29}$$

to which we add the equation $\dot{z} = -u_M/(L/D)$ for the altitude rate, where $L/D$ is the *lift-to-drag ratio*. The similarity of (3.29) to (3.15) is obvious. As shown in Sec. 3.3.4(b), the behaviour of the glide path landing system depends on the parameter $|K \sin\delta|$, $K \triangleq u_M/v_W$ now being called the *wind penetration parameter*. If $|K \sin\delta| < 1$, then the system attains a stable angular alignment as $r \to 0$; if $|K \sin\delta| > 1$, cyclic or orbital paths can occur. The cases are respectively called "target seeking" and "orbital" [ibid.].

For the special case $\delta = 0$, a quasi-pure pursuit, the equation of the trajectory obtained by solving (3.29) is

$$r(\theta) = \frac{r_0}{F(\theta_0)} F(\theta)$$

---

[6]The 'airspeed' of an airborne object is its speed relative to the atmosphere. The 'groundspeed' is its speed relative to the surface of the earth.

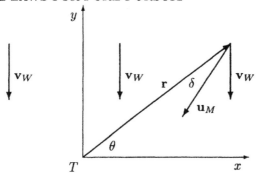

Figure 3.19: Illustration for Example 2, Sec. 3.3.5

where $r_0$ and $\theta_0$ are the initial conditions $(\theta_0 \neq \pm 90^o)$ and

$$F(\theta) \triangleq \frac{[\cos(\theta/2) + \sin(\theta/2)]^{K-1}}{[\cos(\theta/2) - \sin(\theta/2)]^{K+1}} , \quad K > 1. \tag{3.30}$$

This, of course, is another form of the solution (3.7) shown above.

### Example 3 - Navigation along a Rhumb Line

A spherical equivalent of the 'night-insect flight' is the *rhumb line*, the path of a ship that sails on the globe at a constant course $\delta$. (The *course* is defined as the angle between the velocity vector and the local meridian). The projection of this trajectory, which is also called *loxodrome*, from the globe pole on the equator plane is the logarithmic spiral (3.26), $\lambda$ signifying the longitude.[7]

## 3.4   Guidance Laws for Pure Pursuit

The basic idea must already be familiar to the reader: Define an 'error' that quantifies the way the state of the guided object M differs from what it should be according to the geometrical rule, and apply control to reduce it.

The most obvious guidance law for ordinary (i.e., $\delta = 0$) pure pursuit would have the control proportional to the angle between $\mathbf{v}_M$ and $\mathbf{r}$, in the direction of $(\mathbf{v}_M \times \mathbf{r}) \times \mathbf{v}_M$. Alternatively, the (vector) error $\mathbf{e}$ would be proportional to the *cross range* $\mathbf{r}_\perp = (\mathbf{1}_{v_M} \times \mathbf{r}) \times \mathbf{1}_{v_M}$, which is the component of $\mathbf{r}$ across the flight line vector $\mathbf{v}_M$ — and also across the axis of M if the angle of attack $\alpha$ is negligible. ($\alpha$ is the angle between $\mathbf{v}_M$ and the axis of M; see Appendix C.3).

---

[7]The equation of a loxodrome is $\lambda = (\tan \delta) \log \tan(\Lambda/2 + \pi/4)$, where $\Lambda$ is latitude and $\lambda$ longitude. Loxodromes are straight lines on marine charts drawn according to the Mercator projection.

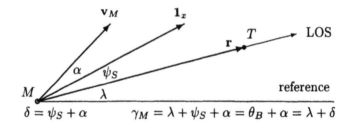

Figure 3.20: Angles involved in velocity- and attitude-pursuits

This law involves some difficulties, which will be dealt with in the following subsections in terms of planar PP.

## 3.4.1   Velocity Pursuit vs. Attitude Pursuit

In defining PP, the direction of $\mathbf{v}_M$ was considered when angles were involved; see e.g. Figs. 3.3 and 3.14 and Equation (3.4). In practice, however, it is advantageous to refer to M's longitudinal body axis vector $\mathbf{1}_x$ rather than to $\mathbf{v}_M$, since LOS angles are most economically measured by instrumentation mounted on the body proper. We recall that aircraft, missiles, and some boats maneuver in such a way that $\mathbf{v}_M$ does not in general coincide with $\mathbf{1}_x$. The angle between $\mathbf{1}_x$ and $\mathbf{v}_M$ is the *angle of attack* $\alpha$ when it is measured about the body $y$-axis. In order to maneuver by aerodynamic (or hydrodynamic, as the case may be) forces, $\alpha$ must be positive (see also Appendix C.3).

(a) *Velocity pursuit* requires $\mathbf{v}_M$ to coincide with the LOS $\mathbf{r}$; this in fact is the geometrical rule we have been studying so far. The guidance law for implementing it would be

$$a_{M_c} = -f_1(\delta) = -f_1(\gamma_M - \lambda) \tag{3.31}$$

(see Fig. 3.20), where $f_1(.)$ is an odd-symmetric function.

In order to mechanize velocity pursuit, the seeker is mounted on a gimbal which weathercocks with the wind, so that it measures $\psi_S + \alpha = \delta$. The idea is not new; in fact it dates from the days of the German Hs-293D television guided bomb mentioned above in Sec. 3.2 [29]. Fig. 3.21 clearly shows the 'weathercock' on which the TV camera is mounted.

(b) *Attitude pursuit* is the geometrical rule that requires $\mathbf{1}_x$ to coincide with $\mathbf{r}$. The guidance law for implementing it would be

$$a_{Mc} = -f_2(\psi_S) = -f_2(\gamma_M - \lambda - \alpha) \tag{3.32}$$

(see Fig. 3.20), where $f_2(.)$ is an odd-symmetric function.

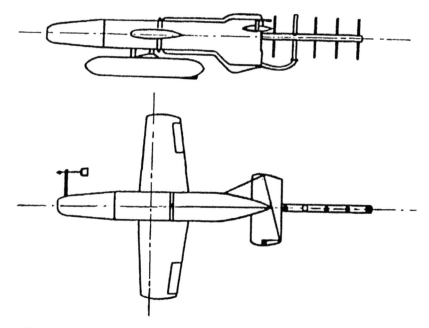

Figure 3.21: Weathercock-mounted TV camera on the Hs-293D (Source: Reference 19, with kind permission from AGARD/NATO)

Attitude pursuit guidance is mechanized by mounting the seeker on the body of M such that the angle it measures is $\psi_S$.

(c) It has been shown that velocity pursuit produces more accurate guidance than attitude pursuit [31, 32]. In an example, Fig. 3.22, adapted from reference [31], it can be seen that miss distances attained by the former are about a third of those attained by the latter.

## 3.4.2 A Simple Velocity-Pursuit Guidance Law

We present here the very simple case where T is stationary. $v_M$ is assumed constant (we shall omit the 'M' subscripts from $v_M$ in this subsection and in the next one). The 'error' in the sense mentioned above is $\delta$, and we recall that

$$\delta = \gamma_M - \lambda .$$ (3.33)

The kinematics equations are

$$\begin{cases} \dot{r} = -v \cos \delta \\ r\dot{\lambda} = -v \sin \delta . \end{cases}$$ (3.34)

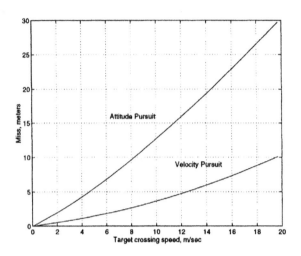

Figure 3.22: Miss distances in velocity- and attitude pursuits

Let the guidance law be (3.31) with $f_1(.) = \sin(.)$, i.e.,

$$\dot{\gamma}_M = -kv\sin\delta \,, \tag{3.35}$$

where $k$ is a constant. $\dot{\lambda} = -(v/r)\sin\delta$ by (3.34), and $\dot{\lambda} = \dot{\gamma}_M - \dot{\delta}$ by (3.33); combining these equations with (3.35) one gets the differential equation

$$r\frac{d\delta}{dr} = (kr - 1)\tan\delta \,, \tag{3.36}$$

whose solution is

$$\frac{\sin\delta}{\sin\delta_0} = \frac{r_0}{r}\epsilon^{k(r-r_0)} \,. \tag{3.37}$$

For $kr_0 > 1$, the ratio $(\sin\delta)/\sin\delta_0$ decreases with $r$ until $r = 1/k$; then it starts increasing until finally, at a certain value of $r$, $\delta$ exceeds the limit imposed by the seeker.

A possible remedy for this problem is to introduce a weighting function of $r$, say $1/r$, into the guidance law, which thus becomes

$$\dot{\gamma}_M = -\frac{kv}{r}\sin\delta \,. \tag{3.38}$$

The differential equation changes to

$$r\frac{d\delta}{dr} = (k - 1)\tan\delta \,, \tag{3.39}$$

and its solution is

$$\frac{\sin \delta}{\sin \delta_0} = \left(\frac{r}{r_0}\right)^{k-1}. \tag{3.40}$$

Hence, with this guidance law $\delta \to 0$ as $r \to 0$ provided $k > 1$. However, since $a_M$ is given by the equation

$$a_M = v\dot{\gamma}_M = -v\frac{kv}{r}\sin\delta = -\frac{kv^2\sin\delta_0}{r_0}\left(\frac{r}{r_0}\right)^{k-2},$$

it follows that in order that $a_M \to 0$ as $r \to 0$, the inequality $k > 2$ must be satisfied.

Guidance law (3.38), stated in terms of curvature rather than $\dot{\gamma}_M$, has been studied for path tracking done by *mobile robots* [33].

A similar law, stated in terms of the lift $L_M$ rather than $\dot{\gamma}_M$ (see Appendix C for the definition of *lift*) is "a basic form of guidance" for missiles designed to *intercept re-entry vehicles* [34]. By this law, the command $L_{M_c}$ is directly proportional to the angle between $\mathbf{v}$ and $\mathbf{r}_{IP}$ and to $v^2$, and inversely proportional to $r_{IP}$, the latter being the range between the interceptor and the *projected intercept point* (PIP). The lift must be directed along the vector $(\mathbf{v} \times \mathbf{r}_{IP}) \times \mathbf{v}$, which is probably why this guidance law is referred to as *cross-product steering* [ibid.]. Ideally, this law brings $\mathbf{v}$ into collinearity with $\mathbf{r}_{IP}$ at the PIP.

### 3.4.3  A Simple Attitude-Pursuit Guidance Law

We make the same assumptions as in the previous subsection, adding the approximations (cf. (3.34))

$$\dot{r} = -v , \quad \dot{\lambda} = -\frac{v}{r}(\psi_S + \alpha) , \tag{3.41}$$

which are valid for $|\delta| = |\psi_S + \alpha| \ll 1$. We also recall the linear relation between $\dot{\gamma}_M$ and $\alpha$,

$$\dot{\gamma}_M = \frac{1}{T_\alpha}\alpha , \tag{3.42}$$

where $T_\alpha$ is the aerodynamic *turning rate time-constant* of M (see Appendix C).

Let the guidance law be (3.32) with $f_2(.)$ linear, i.e.,

$$\dot{\gamma}_M = -kv\psi_S . \tag{3.43}$$

Then, the differential equation

$$\frac{d\psi_S}{\psi_S dr} = \frac{kv}{1 - kvT_\alpha} - \frac{1}{r} \tag{3.44}$$

is derived from the equation $\gamma_M = \lambda + \psi_S + \alpha$ and (3.41)-(3.43). Its solution for $kvT_\alpha < 1$, namely

$$\frac{\psi_S}{\psi_{S0}} = \frac{r_0}{r}\epsilon^{\frac{k}{1-kvT_\alpha}(r-r_0)} , \tag{3.45}$$

is similar to (3.37), $\psi_S$ substituting for $\sin \delta$ and $k/(1 - kvT_\alpha)$, for $k$. It is unsatisfactory for similar reasons.

Changing the guidance law by adding an appropriate weighting function of $r$ can lead to good results, though inferior, in general, to those obtainable by velocity pursuit.

## 3.5   On the Mechanization of Pursuit Guidance

The problems involved with the mechanization of pursuit guidance — pure pursuit or deviated one, velocity pursuit or attitude pursuit — are rather similar to the analogous problems involved with LOS guidance, mentioned in Sec. 2.5.3. Although we consider their treatment to be beyond the scope of this text, the reader is referred to Chapter 6 of this text, which deals with the mechnization of Proportiona Navigation guidance systems, and to the references given there.

## 3.6   References

[1] Yuan, Luke Chia-Liu, "Homing and Navigational Courses of Automatic Target-Seeking Devices", *RCA Laboratories*, Princeton, N.J., Report No. PTR-12C, Dec. 1943.

[2] Yuan, Luke Chia-Liu, "Homing and Navigational Courses of Automatic Target-Seeking Devices", *J. Applied Physics*, Vol. 19, 1948, pp. 1122-1128.

[3] Locke, Arthur S., *Guidance*, a volume in the series *Principles of Guided Missiles Design* edited by Grayson Merrill, Van Nostrand, 1955, Ch. 12.

[4] Russian Translation of [3], Moscow, 1958, translator's footnote on p. 497.

[5] Clarinval, André, *Esquisse historique de la courbe de poursuite*,
UNESCO—Archives internationales d'histoire des sciences, Vol. 10(38), 1957, pp. 25-37.

[6] Bernhart, Arthur, "Curves of Pursuit", *Scripta Mathematica*, Vol. 20, 1954, pp. 125-141.

[7] Bernhart, Arthur, "Curves of Pursuit - 2", *Scripta Mathematica*, Vol. 23, 1957, pp. 49-65.

[8] Bernhart, Arthur, "Curves of General Pursuit", *Scripta Mathematica*, Vol. 24, 1959, pp. 189-206.

[9] Bernhart, Arthur, "Polygons of Pursuit", *Scripta Mathematica*, Vol. 24, 1959, pp. 23-50.

[10] Good, I. J., "Pursuit Curves and Mathematical Art", *Mathematical Gazette*, Vol. 43, 1959, pp. 34-35.

[11] Gardner, Martin, "Mathematical Recreations", *Scientific American*, July 1965.

[12] Bruckstein, Alfred M., "Why the Ants Trails Look so Straight and Nice", *The Mathematical Intelligencer*, Vol. 15, No. 2, 1993, pp. 59-62.

[13] Gabriel, Fritz, "Mathematik der Hundekurve", *Luftwissen*, Bd. 9, No. 1, 1942, pp 21-24.

[14] Mayall, N. U., "Pursuit Curve Characteristics", *Carnegie Institution of Washington, Mount Wilson Observatory*, April 1944 (declassified in Feb. 1948).

[15] Macmillan, R. H., "Curves of Pursuit", *Mathematical Gazette*, Vol. 40, No. 331, Feb. 1956, pp. 1-4.

[16] Eades, James B. and George H Wyatt, "A study from Kinematics—The Problems of Intercept and Pursuit", *Goddard Space Flight Center*, 1969, NASA TMX 63527, N69-23785.

[17] Kohlas, John, "Simulation of Aerial Combats", *Royal Aircraft Establishment (RAE)*, Translation No. 1367, 1969, N71-15372.

[18] Benecke, Th. and A. W. Quick (eds.), *History of German Guided Missile Development*, AGARDograph AG-20, AGARD/NATO (Advisory Group for Aerospace Research and Development, North Atlantic Treaty Organisation) First Guided Missile Seminar, Munich, April 1956.

[19] Münster, Fritz, "A Guiding System Using Television", in [18], pp. 135-160.

[20] Glines, C. V., "The Pigeon Project", *AIR FORCE Magazine*, Feb. 1992, pp. 76-79.

[21] Hölsken, Dieter, *V-Missiles of the Third Reich, The V-1 and V-2*, Sturbridge, Massachusetts, Monogram Aviation Publications, 1994.

[22] Ordway, Frederick and Ronald C. Wakeford, *International Missile and Spacecraft Guide*, McGraw-Hill, 1960.

[23] Güllner, Georg, "Summary of the Development of High-Frequency Homing Devices", in [18], pp. 162-172.

[24] Kutzcher, Edgar W., "The Physical and Technical Development of of Infrared Homing Devices", in [18], pp. 201-217.

[25] Müller, Ferdinand, *Leitfaden der Fernlenkung*, Garmisch-Partenkirschen, Deutsche RADAR, 1955.

[26] Gomes-Teixeira Francisco, "Traité des Courbes Spéciales Remarquables", Paris, Jacques Gabay, 1995 (A reimpression of the first edition, Coïmbre, Académie Royale des Sciences de Madrid, 1908-1915).

[27] Bacon, R. H., "The Pursuit Curve", *J. Applied Physics*, Vol. 21, October 1950, pp. 1065-1066.

[28] Davis, Harold T., *Introduction to Nonlinear Differential and Integral Equations*, New York, Dover, 1962, pp. 113-127.

[29] Fischel, Edouard, "Contributions to the Guidance of Missiles", in [18], pp. 24-38.

[30] Murphy, A. L., *Azimuth Homing in a Planar Uniform Wind*, Natick, Mass., US Army Natick Laboratories, AD-780015, 1973.

[31] Chadwick, W. R. and C. M. Rose, "A Guidance Law for General Surface Targets", *J. Guidance and Control*, Vol. 6, No. 6, 1983, pp. 526-529.

[32] Chatterji, Gano B. and Meir Pachter, "Modified Velocity Pursuit Guidance Law with Crosswind Correlation for Missiles Against Surface Targets", *AIAA Guidance and Control Conf.*, 1991, pp. 1566-1571.

[33] Ollero, A., A. García-Cerezo, and J. L. Martinez, "Fuzzy Supervisory Path Tracking of Mobile Robots", *Cont. Eng. Practice*, Vol. 2, No.2, 1994, pp. 313-319.

[34] Regan, F. J. and S. M. Anandakrishnan, *Dynamics of Atmospheric Re-Entry*, Washington, D.C., AIAA, 1993, Ch. 9.

[35] Mullins, J., "You can run, but you can't hide...", *New Scientist*, 12 April 1997, p. 20.

# Chapter 4

# Parallel Navigation

## 4.1  Background and Definitions

The *parallel navigation* geometrical rule, also called the *constant bearing* rule, has been known since antiquity, mostly by mariners, hence the 'navigation' part of the term. We shall later explain why we prefer 'parallel navigation' to 'constant bearing' (and also to 'collision course navigation' [1] and 'Intercept' [2, 3], yet other names for the veteran geometrical rule).

According to this rule, the direction of the line-of-sight (LOS) $MT$ is kept constant relative to inertial space, i.e., the LOS is kept parallel to the initial LOS; hence the "parallel" part of the term. In three-dimensional vector terminology (see also Appendix A), the rule is very concisely stated as

$$\mathbf{w} = \mathbf{0} \, , \tag{4.1}$$

where $\mathbf{w}$ is the rate of rotation of the LOS $\mathbf{r}$. An equivalent statement of this geometrical rule is

$$\mathbf{r} \times \dot{\mathbf{r}} = \mathbf{0} \, , \tag{4.2}$$

since $\mathbf{r} \times \dot{\mathbf{r}} = r^2 \mathbf{w}$ by (A.3). It follows from (4.2) that the geometrical rule can be stated in yet another way, namely that $\dot{\mathbf{r}}$ and $\mathbf{r}$ *must be colinear*. We have to add the almost obvious caveat that the scalar product $\mathbf{r} \bullet \dot{\mathbf{r}}$ must be negative, otherwise we might have M *receding* from T according to the parallel navigation geometrical rule rather than approaching it.

**Note.** It follows from (4.2) that guidance according to this geometrical rule is *instantaneously planar*, i.e., the vectors $\mathbf{r}(t), \mathbf{v}_T(t)$, and $\mathbf{v}_M(t)$ are coplanar, although the engagement need not be planar. The proof is quite similar to that shown in Sec. 2.3.4 for LOS guidance.

We now leave, for some time, the general case and the three-dimensional vectors associated with it and limit ourselves to the planar case.

## 4.2   Kinematics of Planar Engagements

$\mathbf{v}_M$, $\mathbf{v}_T$ and $\mathbf{r}$ being on the same (fixed) plane by definition, the parallel navigation geometrical rule can be restated as

$$\dot{\lambda} = 0 \qquad (4.3)$$

(with the additional requirement $\dot{r} < 0$), where $\lambda$ is the angle that the LOS forms with the reference line on the said plane (see Fig. 3.14). By (A.13) and (A.14), the equations for $\dot{r}$ and $\dot{\lambda}$ are as follows.

$$\dot{r} = v_T \cos\theta - v_M \cos\delta \qquad (4.4)$$

$$\dot{\lambda} = \frac{v_T \sin\theta - v_M \sin\delta}{r} . \qquad (4.5)$$

In order that $\dot{\lambda}$ be zero, the equation

$$v_M \sin\delta = v_T \sin\theta \qquad (4.6)$$

must be satisfied; in order that $\dot{r}$ be negative, the inequality

$$v_M \cos\delta > v_T \cos\theta \qquad (4.7)$$

must be satisfied. When both (4.6) and (4.7) are satisfied, *straight-line collision course conditions* (CC conditions, for the sake of brevity) are said to prevail since, if $\mathbf{v}_T$ remains constant, M will collide with T by moving on a straight line at the constant velocity $v_M$. Clearly, from everyday experience we know that collision courses need not be straight lines if T changes its speed or direction.

We have already encountered CC conditions in this book. In Sec. 2.3.1(a) it has been shown that these conditions exist just before LOS-guided M intercepts T. In Sec. 3.3.4(b), CC conditions have been shown to result when the (constant) lead angle $\delta$ has the 'optimal' value of $\arcsin[(1/K) \sin\theta]$.

### 4.2.1   Nonmaneuvering Target

(a) If T is not maneuvering, then $v_T$ and $\gamma_T$ are constant. Assuming that $K = v_M/v_T$ is also constant, (4.6) and (4.7) change to

$$\sin\delta = \frac{\sin\theta}{K} \qquad (4.8)$$

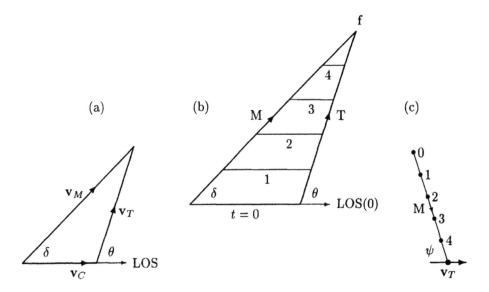

Figure 4.1: Parallel-navigation trajectories, nonmaneuvering target. (a) Velocity collision triangle, (b) trajectories triangle, (c) relative trajectory

$$\cos \delta > \frac{\cos \theta}{K} , \tag{4.9}$$

respectively. When (4.8) and (4.9) are satisfied, CC conditions prevail, and a *collision triangle* can be drawn having the sides $\mathbf{v}_T$, $\mathbf{v}_M$, and $\mathbf{v}_C$ (Fig. 4.1(a)), where $\mathbf{v}_C$ is the *closing speed*, defined as $\mathbf{v}_C \triangleq -\dot{r} = \mathbf{v}_M - \mathbf{v}_T$. When the sides of the triangle are multiplied by the time-of-flight $t_f$, a collision triangle is obtained whose sides are the distances travelled by M and T, respectively, and the initial range $\mathbf{v}_C t_f$ (Fig. 4.1(b)). The trajectory of M 'as seen by T', i.e., in the relative frame of coordinates, is the straight line shown in Fig. 4.1(c). From the point of view of T, M is approaching at a constant speed $v_C$ and at a *constant bearing* angle $\psi = -\theta$ 'off its tail'; hence the alternative name for the geometrical rule. However, we shall see that when T maneuvers, the approach of M is no more on a constant bearing line; this is the reason why 'parallel navigation' is preferred as the name for the present geometrical rule rather than 'constant bearing'.

(b) An application for mariners wishing to rendezvous each other at sea is obvious and has been known for centuries. M could be a boat and T, a tanker with fuel for it (or vice versa). Or, back in history, T could be a merchantman and M a pirate ship. There is also the inverse application, i.e., a rule of thumb for avoiding collisions: Steer away from a situation where another boat is approaching you at

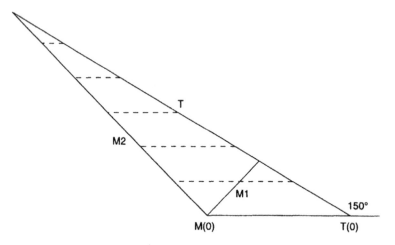

Figure 4.2: Two solutions for a parallel-navigation problem

a constant bearing. This ancient and very good rule assumes, of course, constant speeds.

(c) Sometimes there is no solution to (4.8) and (4.9); for example, for $K \leq 1$ and $|\theta| \leq 90^\circ$. On the other hand, for $K < 1$ and $|\theta| > 90^\circ$ there are two solutions. For example, with $K = 1/\sqrt{2}$ and $\theta = 150^\circ$, the solutions for $\delta$ are $45^\circ$ and $135^\circ$, for which the distances travelled by M are $0.518\, r_0$ and $1.932\, r_0$, respectively, $r_0$ being the initial range. In Fig. 4.2, $M_1$ is the trajectory of M for $\delta = 45^\circ$, and $M_2$ for $\delta = 135^\circ$.

When there is a solution, $r$ varies with time $t$ according to the equation

$$r(t) = \int_0^t \dot{r}(\tau)d\tau = r_0 + \dot{r}t$$

since $\dot{r}$ is constant. (See also the illustration in Fig. 4.1, where time graduation—in arbitrary units—is shown). Total time of flight $t_f$ is obtained from (4.4) by letting $r(t_f) = 0$:

$$t_f = -\frac{r_0}{\dot{r}} = \frac{r_0}{v_T}\, \frac{1}{K \cos \delta - \cos \theta} \ . \tag{4.10}$$

(d) The equations for $t_f$-isochrones are obtained from (4.10) in much the same way that $t_f$-isochrones were obtained for pure-pursuit (PP) from (3.10). It can be shown that these isochrones are now *circles*, as compared with ellipses for the PP case. In fact, for the same values of $K$, the circles are concentric with the ellipses and are tangent to them at $\psi = 0$ and $180^\circ$ (Fig. 4.3). Recall that the circles and

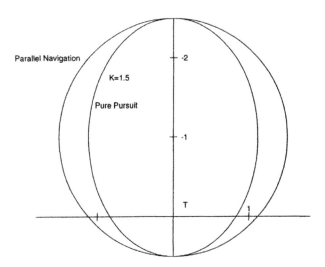

Figure 4.3: $t_f$-isochrones for parallel navigation and pure pursuit

the ellipses are drawn in the TT FOC (this term has been defined in Sec. 3.3.1(b)). The unit of distance is $v_T t_f$.

(e) $t_f$-isochrones in *absolute* coordinates are clearly circles centered at $T$ having the radius $v_T t_f$ and circles centered at $M$ having the radius $v_M t_f$. The circles intersect each other at the *interception point* (IP) for $t_f$. It is shown in Appendix D.6 that the *loci of interception* are two families of circles, for $K > 1$ and $K < 1$, respectively.[1] If $M(0) = (-r_0/2,\ 0)$ and $T(0) = (r_0/2,\ 0)$, then the two families of circles are symmetric about the $y$-axis and are given by the equation

$$\left(\frac{x}{r_0} - \frac{K^2 + 1}{2(K^2 - 1)}\right)^2 + \left(\frac{y}{r_0}\right)^2 = \left(\frac{K}{K^2 - 1}\right)^2, \quad K \neq 1. \tag{4.11}$$

For $K = 1$, the interception locus is obviously the $y$-axis itself. Note that the circle for $1/K$ is the mirror image of the circle for $K$.

For an illustration, interception loci for $K = 2^{1/2}, 2^{3/4}, 2, 2^{5/4}$ and their reciprocals are shown in Fig. 4.4. Also shown in the figure are the triangles of the Example given in (c) (and depicted in Fig. 4.2) and the two isochrones for $t_f = 2.732(r_0/v_T)$ which intersect each other at one of the IP's of that example.

---

[1] Each of the circles is a Circle of Apollonius (named after Apollonius of Perga, 260-200 B.C., founder of conic sections theory), being the locus of a point whose distances from two fixed points are in a constant ratio.

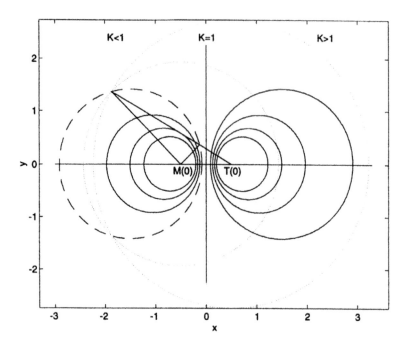

Figure 4.4: Interception loci for several values of $K$ and $1/K$.

## 4.2.2   Maneuvering Target

Parallel navigation is optimal for nonmaneuvering targets: intercept is achieved in minimum time $t_f$, and no maneuver at all is required from M, i.e., $a_M = 0$ throughout the engagement. When T does maneuver, by changing either its speed or its course (or both), parallel navigation may still be quite effective but will not in general provide the 'best' trajectories for M. In other words, the acceleration $a_M$ will differ from 0 most of the time and, furthermore, M's trajectory will not be the shortest possible. We shall examine two examples.

### Example 1 - T Moving on a Circle

For the first example, T is assumed to move on a circle with radius $c$ with constant angular velocity $\omega = \dot{\gamma}_T = v_T/c$. The equations for the kinematics of this case are

$$\begin{cases} \dot{r} = v_T \cos\theta - v_M \cos\delta \\ \dot{\lambda} = \frac{1}{r}(v_T \sin\theta - v_M \sin\delta) = 0 \\ \dot{\theta} = \dot{\gamma}_T - \dot{\lambda} = \omega \\ a_M = v_M\dot{\gamma} = v_M\dot{\delta} \ . \end{cases} \qquad (4.12)$$

An important consequence of these equations is the inequality (proven in Appendix D.7)

$$a_M(t) \leq a_T = \omega v_T \text{ for all } t \qquad (4.13)$$

provided $K \geq 1$. This of course is a great advantage of parallel navigation. A more general statement of this remarkable property is the following theorem.

M can capture T from any initial state if and only if $v_M > v_T$ *and* $a_{M_{max}} \geq a_{T_{max}}$, the speeds $v_M$ and $v_T$ being constant, where $a_{(.)_{max}}$ denotes maximum lateral-acceleration capability [4].

Time is not proportional to $r_0 - r$ in the present case, since $\dot{r}$ is not constant. $t_f$ is found from the integral

$$\int_0^{t_f} \dot{r}(\tau)d\tau = r(t_f) - r_0 = -r_0 . \qquad (4.14)$$

It is given in terms of $\theta_f = \theta_0 + \omega t_f$ by the implicit equation (proven in Appendix D.8)

$$K[E(\theta_f, \frac{1}{K}) - E(\theta_0, \frac{1}{K})] - (\sin\theta_f - \sin\theta_0) = \frac{r_0}{c} , \quad K > 1 , \qquad (4.15)$$

where $E(.,1/K)$ is an elliptic integral of the second kind.

For an illustration, parallel navigation trajectories are shown in an absolute FOC, Fig. 4.5(a), where T, say an aircraft, is flying on a circle with radius 1, centered at the origin. Three missiles are launched at it from its right-hand side, from the points $(2.5, 0), (2, 0)$, and $(1.5, 0)$, such that the initial ranges are $1.5, 1.0$ and $0.5$, respectively. The velocity ratio is $K = 2$. Lines-of-sight at 0.25-time intervals are also shown in the drawing, in order to emphasize the fact that we are dealing with the parallel-navigation geometrical rule. (The unit of time is $1/\omega_T$, where $\omega_T$ is the angular velocity of T). It is easily seen that, although all the missiles are launched at aspect angles $\psi = 90^\circ$, they intercept T from smaller angles; in fact, for the missile launched from the farthest point, the final aspect angle is approximately $30^\circ$. In other words, since T is maneuvering, M approaches it from a variable bearing angle.

This property is even better illustrated by the *relative* trajectories of the missiles, shown in Fig. 4.5(b). Note also that the 'distances' covered by M along the relative trajectories during the 0.25-time intervals (the lines-of-sight are here rays that start at the origin) are not constant, although both $v_T$ and $v_M$ are. This of course results from the fact that these trajectories are shown in a relative FOC.

For $K = 1$, M can intercept T only if $\cos\theta < 0$ during at least some part of the engagement and provided $r_0 \leq 4c$; however, these are but necessary conditions.

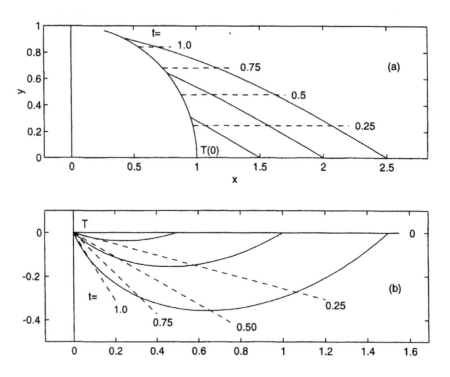

Figure 4.5: Parallel navigation trajectories, maneuvering target, Example 1 in Sec. 4.2.2

**Example 2 - Variable $v_T$**

Suppose now that T maneuvers by changing its speed, keeping $\gamma_T$ constant. We now have the equation

$$\sin \delta(t) = \frac{v_T(t) \sin \theta}{v_M} \tag{4.16}$$

where $\theta$ is constant, and assume that the condition $\dot{r} < 0$ is satisfied. It is seen that

(i) Although M approches T along a straight line in a relative FOC, i.e., at a constant relative bearing, the trajectory of M in an absolute one is *not* a straight line;

(ii) $\dot{r}$ is not constant, and therefore

(iii) $t$ is not proportional to $r_0 - r$.

By differentiating (4.16) with respect to time, bearing in mind that $a_M =$

$v_M \dot\gamma_M = v_M \dot\delta$, the expression

$$a_M = \frac{\dot v_T \sin\theta}{\cos\delta} = \dot v_T \frac{K\sin\theta}{\sqrt{K^2 - \sin^2\theta}} \qquad (4.17)$$

is easily obtained (note that $K$ is not constant here). Whereas $\dot v_T$ is quite low for such targets as aircraft or motorcars, it may attain high values if T is a penetrating ballistic missile, for example.

For an illustration, suppose T is travelling on the line $x_T(t) = c$ with $y_T(0) = 0$ and $\dot y_T(t) = v_T$, reducing its speed according to the equation

$$v_T(t) = \begin{cases} Uf(t), & t \le T \\ 0, & t > T , \end{cases}$$

where $c, U$ and $T$ are constants and $f(T) = 0$. Clearly the trajectory of T is given by the equations

$$x_T(t) = c, \quad y_T(t) = U\int_0^t f(t')dt' .$$

M starts parallel navigation at the origin, having the constant speed $KU$ (here $K$ is again constant). The LOS is parallel to the $x$-axis and the $x$ component of $v_M$ is $KU\cos\delta$. M's trajectory is therefore given by the equations

$$x_M(t) = KU\int_0^t \cos\delta(t')dt', \quad y_M(t) = y_T(t) ,$$

where $\cos\delta(.)$ is found from the equation

$$\sin\delta = \frac{v_T}{v_M}\sin\theta = \frac{v_T}{v_M} = \frac{1}{K}f(t) ,$$

so that

$$x_M(t) = U\int_0^t \sqrt{K^2 - f^2(t')}dt' .$$

The time-of-flight $t_f$ is found from the equation $x_M(t_f) = c$. As $\theta = 90°$, the lateral acceleration required is, by (4.17), $\dot v_T/\cos\delta$, or

$$a_M(t) = \frac{KU\frac{df(t)}{dt}}{\sqrt{K^2 - f^2(t)}} .$$

To be more specific, let us examine three cases, as follows.
(i) $f(t) = 1 - t/T$;
(ii) $f(t) = 1 - (t/T)^2$;
(iii) $f(t) = \sqrt{1 - (t/T)^2}$.
The three respective trajectories are shown in Fig. 4.6(a) for $K = 1.5$. The values of $t_f$ are 0.745, 0.797, and 0.825, respectively, in units of $c/U$. Note that, in spite of the similarity between the trajectories, the graphs for lateral accelarations differ greatly from one another, as is evident from Fig. 4.6(b).

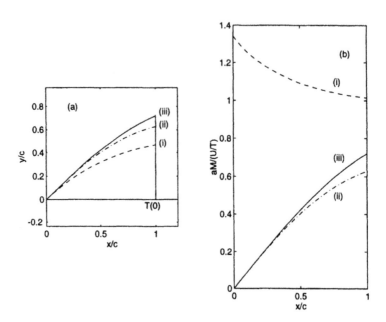

Figure 4.6: Trajectories for three decelerating targets, Example 2 in Sec. 4.2.2

## 4.2.3   Variable Speed

Now let $v_T$ and $\gamma_T$ be constant and $v_M$ variable. This is a very common situation with missiles propelled by solid-fuel rocket motors, where the thrust of the rocket motor contributes to positive $\dot{v}_M$, and aerodynamic drag to negative $\dot{v}_M$. In the present case there is, so to speak, a coupling between the axial and the lateral accelerations, given by the equation

$$\frac{a_M}{v_M} = -\tan\delta = -\frac{\sin\theta}{\sqrt{K^2 - \sin^2\theta}} \ , \tag{4.18}$$

obtained in the same way (4.17) was, $K$ again not being constant. For an illustration of the effects of this coupling, see the example given in Sec. 6.4.1(a), where parallel navigation is examined along with 'proportional navigation'.

(4.17) and (4.18) can in fact be combined, yielding the equation

$$a_M = (K\dot{v}_T - \dot{v}_M)\tan\delta \ . \tag{4.19}$$

Note that, disregarding the sign, the ratio $a_M/\dot{v}_T$ is $K$ times the ratio $a_M/\dot{v}_M$. However, $\dot{v}_M$ is usually quite high compared to $\dot{v}_T$.

We sum up the examples of this subsection by observing that, in all of them, parallel navigation is seen not to be optimal in the sense it is when both $\mathbf{v}_T$ and $v_M$ are constant.

## 4.3 Nonplanar Engagements

### 4.3.1 Definitions

We first recall the alternative definition (4.2) of parallel navigation, namely $\mathbf{r} \times \dot{\mathbf{r}} = \mathbf{0}$. The requirement $\mathbf{r} \bullet \dot{\mathbf{r}} < 0$ has been added in order to ensure that M should approach T not recede from it. (4.2) can also be stated as

$$\mathbf{r} \times (\mathbf{v}_T - \mathbf{v}_M) = \mathbf{0} \ ,$$

or, equivalently, as

$$\mathbf{1}_r \times \mathbf{v}_M = \mathbf{1}_r \times \mathbf{v}_T \ .$$

When multiplied (a vector product) by $\mathbf{1}_r$, the left hand side of this equation becomes $\mathbf{v}_{M_\perp}$, and the right hand side $\mathbf{v}_{T_\perp}$, where $\mathbf{v}_{M_\perp}$ and $\mathbf{v}_{T_\perp}$ are the components of $\mathbf{v}_M$ and $\mathbf{v}_T$ across the LOS $\mathbf{r}$, respectively (see (A.2) in Appendix A.1). Hence one has the equation

$$\mathbf{v}_{M_\perp} = \mathbf{v}_{T_\perp} \ , \tag{4.20}$$

or, stated differently,

$$\mathbf{v}_{C_\perp} = \mathbf{0} \ , \tag{4.21}$$

where $\mathbf{v}_{C_\perp}$ is the component of the closing velocity $\mathbf{v}_C$ across the LOS. In planar terms this was stated as (4.6).

Similarly, the condition $\mathbf{r} \bullet \dot{\mathbf{r}} < 0$ leads to the inequality

$$\mathbf{1}_r \bullet \mathbf{v}_M > \mathbf{1}_r \bullet \mathbf{v}_T \ ,$$

i.e., the component of $\mathbf{v}_M$ *along* the LOS should be greater than the component of $\mathbf{v}_T$ along it. The planar equivalent is (4.7).

### 4.3.2 Three properties

**(a)** It results from (4.20) and (A.10) that $\mathbf{r}$, $\mathbf{v}_T$ and $\mathbf{v}_M$ are coplanar, i.e., a parallel navigation engagement is *instantaneously* planar. Furthermore, if $\mathbf{v}_T$ is constant, then the engagement is planar, i.e., takes place on a *fixed* plane.

**(b)** If both $\mathbf{v}_T$ and $v_M$ are constant, then so are $\mathbf{v}_M$ and $\dot{r}$ and, consequently, $\dot{r}$. Time is therefore related to range by the linear equation $t = (r_0 - r)/(-\dot{r})$. $t_f$ is of course $-r_0/\dot{r}$.

(c) It has been shown that inequality (4.13), namely $a_M(t) \leq a_T(t)$, is true for the general, three-dimensional case as well as for the planar one [5]. A more general statement of this property is the following theorem.

A sufficient condition for intercept is

$$
\begin{cases}
v_M > v_T \\
a_{M_{max}} > a_{T_{max}}
\end{cases}
\tag{4.22}
$$

provided both $v_M$ and $v_T$ are constant [6]. The similar condition for the planar engagement case, mentioned above, requires just "$\geq$" in the second inequality, and is sufficient *and* necessary.

## 4.3.3  Examples

### Example 1 - Nonmaneuvering T

In this example, the engagement is planar, $\mathbf{v}_T$ and $\mathbf{v}_M$ are constant and, consequently, the trajectory is a straight line.

(a) Suppose T is an aircraft approaching the origin of a Cartesian FOC $(x, y, z)$. At $t = 0$, T is at the point $(x_0, y_0, z_0)$ and M starts parallel navigation guidance from the origin so as to intercept it. The velocity of T is $U$ (constant), in the $-x$ direction, and the velocity of M is $KU$, also constant (Fig. 4.7). We wish to find the flight direction of M (determined by the components of $\mathbf{v}_M$), the flight time $t_f$, and the lead angle $\delta$.

We start by finding $\theta$, using the equation

$$
\cos \theta = \frac{\mathbf{r}_0 \bullet \mathbf{v}_T}{r_0 v_T} .
$$

$\delta$ is then found by (4.8). The next step is to obtain the three components of $\mathbf{v}_M$. One equation results from the scalar product

$$
\mathbf{r}_0 \bullet \mathbf{v}_M = r_0 v_M \cos \delta .
$$

The two other equations are obtained from (4.2):

$$
\mathbf{r} \times \dot{\mathbf{r}} =
\begin{vmatrix}
\mathbf{1}_x & x_0 & -U - v_{M_x} \\
\mathbf{1}_y & y_0 & -v_{M_y} \\
\mathbf{1}_z & z_0 & -v_{M_z}
\end{vmatrix}
= \mathbf{0} .
\tag{4.23}
$$

Let us now introduce numerical values. $x_0$, $y_0$ $z_0$ are 12, 3, and 4 $km$, respectively, hence $r_0 = \sqrt{12^2 + 3^2 + 4^2} = 13\,km$, $U = 250\,m/sec$, and $K = 2$.

The equation for $\theta$ gives

$$
\theta = \arccos \frac{12 \times (-250)}{13 \times 250} = 157.4° ,
$$

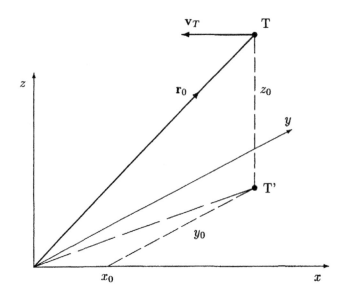

Figure 4.7: Geometry of Example 1 in Sec. 4.3.3

hence

$$\delta = \arcsin \frac{250}{2 \times 250} \sin \theta = 11.1^\circ .$$

The first equation for the components of $\mathbf{v}_M$ is therefore

$$12 v_{M_x} + 3 v_{M_y} + 4 v_{M_z} = 13 \times 500 \times \cos \delta = 6378.7 \; km.m/sec.$$

The equations obtained from (4.23) are

$$\begin{cases} z_0 v_{M_y} - y_0 v_{M_z} = 0 \\ y_0 v_{M_x} - x_0 v_{M_y} = -y_0 U . \end{cases} \tag{4.24}$$

Thus one finally has the set of three equations

$$\begin{cases} 12 v_{M_x} + 3 v_{M_y} + 4 v_{M_z} = 6378.7 \\ 4 v_{M_y} - 3 v_{M_z} = 0 \\ 3 v_{M_x} - 12 v_{M_y} = -750 , \end{cases} \tag{4.25}$$

whose solution is

$$\begin{cases} v_{M_x} = 1.664 \, U \\ v_{M_y} = 0.666 \, U \\ v_{M_z} = 0.888 \, U . \end{cases} \tag{4.26}$$

The given $\mathbf{r}_0$ and $\mathbf{v}_T$ define the *guidance plane*, on which the parallel navigation engagement takes place; $\mathbf{v}_M$ is also on this plane. $\delta$ and $\theta$, already found, are measured about axes normal to this plane.

The duration of flight is now found by (4.10) to be

$$t_f = \frac{13000}{-(250\cos\theta - 500\cos\delta)} = 18.0 \; sec.$$

**(b)** We can use this example to illustrate an application of Euler-angle transformations (see also Example 2 in Sec. 2.3.4; Appendix B is a brief reminder for this useful technique).

It is sometimes desired to do certain guidance calculations in a Cartesian FOC such that (i) $\mathbf{r}_0$ is along the $x$-axis, and (ii) the guidance plane is $z = 0$, such that $\mathbf{v}_T$ has only the components $v_{T_x}$ and $v_{T_y}$, along and across $\mathbf{r}$, respectively. One way of doing it is as follows (subscripts "0" will be omitted from $\mathbf{r}_0$ and its components).

First, rotate the original FOC $(x, y, z)$ by an angle $\psi_1$ about the $z$-axis (see Figs. 4.7 and 2.13). If $\psi_1$ is chosen to be $\arctan(y/x) = \arctan(3/12) = 14.04°$, then the components of $\mathbf{r}$ in the new FOC $(x', y', z')$ are such that $y' = 0$ and $z' = z$. Next, rotate the $(x', y', z')$ FOC about the $y'$-axis by an angle $\theta_2$; if this angle is chosen to be $-\arcsin(z/r) = -\arcsin(4/13) = -17.92°$, then the components of $\mathbf{r}$ in the new FOC $(x'', y'', z'')$ will be $(r, 0, 0) = (13, 0, 0)$.

These two rotations are best described by the matrix product

$$\begin{bmatrix} x'' \\ y'' \\ z'' \end{bmatrix} = \begin{pmatrix} c\theta_2 & 0 & -s\theta_2 \\ 0 & 1 & 0 \\ s\theta_2 & 0 & c\theta_2 \end{pmatrix} \begin{pmatrix} c\psi_1 & s\psi_1 & 0 \\ -s\psi_1 & c\psi_1 & 0 \\ 0 & 0 & 1 \end{pmatrix} \begin{bmatrix} 12 \\ 3 \\ 4 \end{bmatrix} = \begin{bmatrix} 13 \\ 0 \\ 0 \end{bmatrix} \qquad (4.27)$$

where $c$ and $s$ denote cos and sin, respectively.

A third rotation is about the $x''$-axis, by an angle $\phi_3$. $\mathbf{v}_T$ in the resulting FOC is given by the equation

$$\begin{bmatrix} v_{T_x'''} \\ v_{T_y'''} \\ v_{T_z'''} \end{bmatrix} = \begin{pmatrix} 1 & 0 & 0 \\ 0 & c\phi_3 & s\phi_3 \\ 0 & -s\phi_3 & c\phi_3 \end{pmatrix} \mathbf{T}_y(\theta_2)\mathbf{T}_z(\psi_1) \begin{bmatrix} -U \\ 0 \\ 0 \end{bmatrix} , \qquad (4.28)$$

where $\mathbf{T}_y(\theta_2)$ and $\mathbf{T}_z(\psi_1)$ are the respective matrices shown in (4.27). From (4.28) one gets (see also (B.7))

$$\begin{bmatrix} v_{T_x'''} \\ v_{T_y'''} \\ v_{T_z'''} \end{bmatrix} = \begin{bmatrix} -c\theta_2 c\psi_1 \\ c\phi_3 s\psi_1 - s\phi_3 s\theta_2 c\psi_1 \\ -s\phi_3 s\psi_1 - c\phi_3 s\theta_2 c\psi_1 \end{bmatrix} U . \qquad (4.29)$$

By (4.29), if $\phi_3 = -\arctan(\sin\theta_2 \cot\psi_1) = \arctan(zx/ry) = 50.91°$, then $v_{T_{x'''}} = 0$ as desired, and

$$\begin{cases} v_{T_{x'''}} = -U\cos\psi_1\cos\theta_2 = -0.923\,U \\ v_{T_{y'''}} = (\cos\phi_3\sin\psi_1 - \sin\phi_3\sin\theta_2\cos\psi_1)U = 0.385\,U \ . \end{cases} \tag{4.30}$$

For a check we note that $\arccos(-0.923) = \arcsin(0.385) = 157.4°$, as obtained before.

### Example 2 - T Moving on a Circle, M on a Cylinder

This is an example for a simple nonplanar engagement.

Suppose T moves circularly on the plane $z = h$ according to the parametric equations

$$x(t) = c\cos\omega t, \quad y(t) = c\sin\omega t \ ,$$

its velocity $v_T = \omega c$ being constant. At $t = 0$, from the point $(c, 0, 0)$, M starts parallel navigation towards T, its velocity $v_M = K v_T$ being constant, too. $K > 1$ is assumed.

It follows from the definition of the engagement that the LOS is parallel to the $z$-axis throughout the engagement and that $\mathbf{v}_M$ leads it by the angle $\delta = \arcsin(1/K)$. Clearly, the engagement is not planar although, by Property (a) of Sec. 4.3.2, it is always instantaneously planar. In fact, the trajectory of M is a helix, or what mechanical engineers would call "a constant-pitch screw thread on the envelope of a cylinder with a radius $c$", symmetrical about the $z$-axis. (It may be of interest to compare this example with Example 1 of Sec. 2.3.4).

### Example 3 - T Moving on a Circle

Here we wish to present a nonplanar case where M's trajectory is neither a straight line as in Example 1 nor a simple spatial curve as is in Example 2.

T's motion is the same as in Example 2, i.e., it traces a circle with radius $c$ on the plane $z = h$, with $x(t) = c\cos\omega t$ and $y(t) = c\sin\omega t$. At $t = 0$, M starts parallel navigation at the point $(x_{M_0}, y_{M_0}, z_{M_0})$ with $z_{M_0} \neq h$. Clearly this is a nonplanar engagement; furthermore, one has to resort to numerical calculation in order to find M's trajectory.

The procedure of the solution is somewhat similar to that of Example 1. First one finds the aspect angle $\theta$ from the equation $\cos\theta = \mathbf{v}_T \bullet \mathbf{r}/v_T r$ and the lead angle $\delta$ from $\sin\delta = (1/K)\sin\theta$. Then (4.20) is used, yielding

$$\mathbf{v}_{M_\perp} = \mathbf{v}_{T_\perp} = \mathbf{1}_r \times (\mathbf{v}_T \times \mathbf{1}_r) \ .$$

Now the component of $\mathbf{v}_M$ along the LOS is

$$\mathbf{v}_{M_\parallel} = \mathbf{1}_r v_M \cos\delta \ ,$$

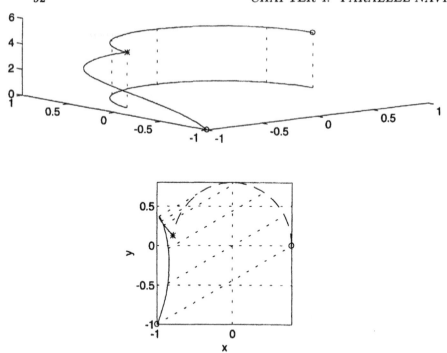

Figure 4.8: Trajectories, Example 3 in Sec. 4.3.3

so that, finally, $\mathbf{v}_M$ is given by the sum $\mathbf{v}_{M_\perp} + \mathbf{v}_{M_\parallel}$. The trajectory is found by integration, i.e., $\mathbf{r}_M(t) = \int_0^t \mathbf{v}_M(t')dt'$.

An illustration is shown in Fig. 4.8 for which $v_T = 200\,m/s$, $h = 4300\,m$, $c = 800\,m$, $K = 2$, and $x_{M_0} = y_{M_0} = -1000\,m$. In the upper part of the figure, a three-dimensional view is shown of M's and T's trajectories, as well as the projection of the latter on the horizontal, as it were, plane $z = 0$. In the lower part, the projection of both trajectories on the same plane is shown, along with the projection of the lines of sight at equal time intervals.

## 4.4    Guidance Laws for Parallel Navigation

### 4.4.1    Proportional Navigation

We recall that the parallel navigation geometrical rule has been defined as $\mathbf{w} = \mathbf{0}$ for the general case and $\dot{\lambda} = 0$ for the planar one. Therefore, following the same logic as in the previous chapters, the obvious choice for a guidance law would be to define the 'error' $e$ as $\mathbf{w}$ (or $\dot{\lambda}$) and make the control $\dot{\gamma}_{M_c}$ or $a_{M_c}$ be proportional to $e$ or to an odd function of it.

Indeed, this is how the guidance law

$$\dot{\gamma}_{M_c} = N\dot{\lambda} \ , \tag{4.31}$$

known as *proportional navigation*, was probably born. An equivalent form of this law is

$$a_{M_c} = Nv_M\dot{\lambda} \ . \tag{4.32}$$

N is called the *navigation constant*. In the general, nonplanar formulation, one form of the law is

$$\mathbf{a}_{M_c} = N\mathbf{w} \times \mathbf{v}_M \ , \tag{4.33}$$

other forms also having $\mathbf{w}$ in a vector product.

Proportional navigation (PN), being very wide-spread in military and other applications on one hand, and having many analytical problems involved with it on the other hand, requires a separate chapter. In fact, the next three chapters of this book are devoted to proportional navigation and related topics.

### 4.4.2 A Non-Feedback Law

(a) M cannot in general predict the motions of T. It can, however, assume that as of the moment $t$ until the presumed interception at $t_f$, T does not maneuver, so that $\mathbf{v}_T$ remains constant.

If the assumption is correct, and if M chooses the lead angle $\delta_{id}$ required by the parallel navigation geometrical rule, namely $\arcsin[(1/K)\sin\theta]$, then interception occurs; if M chooses a 'wrong' lead angle, i.e., $\delta \neq \delta_{id}$, then it misses T. It can be shown that the miss distance $m$ that results is proportional to the initial range $r_0$ and to $v_{C_\perp}$, the component of $\mathbf{v}_C$ perpendicular to $\mathbf{r}_0$:

$$m = r_0\frac{v_{C_\perp}}{v_C} \ . \tag{4.34}$$

Alternatively, $m$ can be expressed in terms of the angle between $\mathbf{v}_C$ and $\mathbf{r}_0$, as follows.

$$m = r_0\sin(\mathbf{v}_C, \mathbf{r}_0) \ . \tag{4.35}$$

See Appendix D.9 for the derivation of (4.34) and (4.35).

If the assumption is incorrect and T starts maneuvering at, say, $t = 0$, then miss will in general result. For an illustration, let $T(0) = (0,0)$, $\mathbf{v}_T(0) = [v_T, 0]^T$, and $M(0) = (0, -\sqrt{3})$, such that $\lambda(0) = 90^o$ and $\theta_0 = -90^o$ (see Fig. 4.9). Let $K = 2$. Clearly, the lead angle dictated by the parralel navigation rule is $-30^o$ and the projected IP is $(1,0)$. Suppose now that T executes a circular maneuver with radius $= 1$ starting at $t = 0$, either clockwise or counterclockwise. The respective trajectories are shown in the figure. The points in time when closest approaches

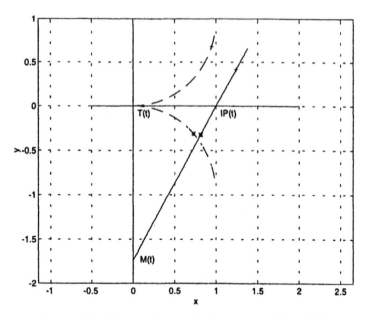

Figure 4.9: Misses that result from unpredicted T's maneuvers

occur are indicated by '+' and '*' signs, respectively. The numerical values of the miss distances are 0.392 and 0.087.

(b) Suppose M makes the assumption stated above about T keeping its velocity $\mathbf{v}_T$ constant and, furthermore, calculates the moment of intercept $t_f$ according to the present conditions. In other words, at any point in time $0 \leq t < t_f$, M expects intercept to occur at $t_f$ such that the projected IP (PIP) is a distance $\mathbf{v}_T \tau$ ahead of T, where $\tau = t_f - t$ is the *time to go*. It follows that, if $t_f$ is kept constant, the locus of IP is an involute[2] of T's path. Equivalently, T's path is the evolute[3] of the trajectory of IP.

Since any involute is orthogonal to the tangents to its evolute, and since T's maneuver $\mathbf{a}_T$ is perpendicular to its velocity $\mathbf{v}_T$ (assuming $v_T$ is constant), it turns out that the interception point IP moves in the same direction as $\mathbf{a}_T$. This fact is said to be of help to aircraft pilots in certain combat situations.

For an illustration, let T's path during its maneuver be the circle $x_T(t) = \sin t$, $y_T(t) = 1 - \cos t$. As T moves on this circle, IP moves on an involute to the circular path, until they meet when $t = t_f$. The kinematics is shown in Fig. 4.10, for which $v_T t_f = \pi/2 \approx 1.57$ has been chosen. A tangent to T's circle is also

---

[2]To obtain an involute of a curve $C$, one takes any fixed point $F$ on $C$, and along the tangent at a variable point $P$ one measures off a length $PQ$ in the directin from $F$, such that $arcFP + PQ = const$. In the present case, $v_T t_f$ is the constant.

[3]An evolute of a curve is the locus of its centers of curvature.

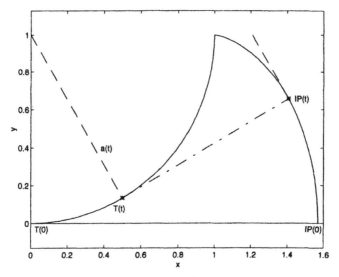

Figure 4.10: T's circular path and a corresponding involute

shown, as well as the corresponding IP and the direction of $\mathbf{a}_T$.

## 4.5 Rules Related to Parallel Navigation

### 4.5.1 Constant Aspect Navigation

In *constant-aspect navigation* (CAN), the (aspect) angles[4] that the LOS forms with the target axis (actually, with $\mathbf{v}_T$) are kept constant [8]. In planar engagements this rule implies

$$\dot\theta = 0 \; . \tag{4.36}$$

In other words, the trajectory of M relative to T is, by definition, a straight line. This geometrical rule is not a mere arbitrary definition. "airplane pilots (...) perform particularly well in (...) tracking tasks with displays that allow them to anticipate and maintain constant angular position relative to a target. Predators and organisms pursuing mates commonly adjust their position to maintain control of relative angle of motion between the pair" [9].

Clearly, for the planar, constant-$v_T$, constant-$v_M$ case, CAN is equivalent to parallel navigation; see e.g. Fig. 4.1. In other cases, M has to maneuver, by

---

[4]The aspect angle $\theta$ (or its negative, $\psi$, see Fig. 3.3) is often referred to as the 'relative bearing' or just 'bearing', and the LOS angle $\lambda$, as the 'absolute bearing' or again just 'bearing'. Therefore, in order to avoid confusion, we prefer the terms 'constant aspect' and 'parallel navigation' for the respective geometrical rules to the ambiguous term 'constant bearing'.

changing either its speed or its path angle (or both).

For an illustration, let us examine the case where T is moving on a circle such that the parametric equations of its trajectory in the $(x, y)$ plane are

$$x_T(t) = c \cos \phi(t), \quad y_T(t) = c \sin \phi(t), \quad \phi(t) = \omega t \ ,$$

$c$ and $\omega$ being constant. M intercepts it following the CAN geometrical rule such that $\psi \equiv -\theta = \pi/2$, either (a) at the constant speed $v_M$ or (b) varying $v_M$ so as to keep the closing speed $v_C$ constant. The value of $\pi/2$ for $\psi$ is arbitrary, chosen for convenience; any other angle may be chosen.

In either case, it follows from the definition of CAN and the geometry of the example that M's trajectory in the $(x, y)$ plane is given by the equations

$$x_M(t) = r(t) \cos \phi(t), \quad y_M(t) = r(t) \sin \phi(t) \tag{4.37}$$

where $r$, of course, is the distance of M from the origin.

In Case (a), where $v_M = const.$, it follows from the kinematics equation

$$\dot{r}^2 + (r\dot{\phi})^2 = v_M^2$$

(cf. (1.6)) that

$$r(t) = Kc \sin(\omega t + \arcsin \frac{r_0}{Kc}) \tag{4.38}$$

where $K = v_M/v_T$. Clearly the initial condition must satisfy the inequality $r_0 \leq Kc = v_M/\omega$. It can easily be shown that (4.38) leads to the parametric equations of the circle

$$\begin{cases} x_M(t) = \frac{Kc}{2}[\sin \chi + \sin(2\omega t + \chi)] \\ y_M(t) = \frac{Kc}{2}[\cos \chi - \cos(2\omega t + \chi)] \end{cases}$$

where $\chi \triangleq \arcsin(r_0/Kc)$.

In Case (b), where $v_C = const.$, the trajectory of M is obviously described by the equations (4.37) with $r(t) = r_0 - v_C t$. It turns out to be an *Archimedean spiral*.

The trajectories for both cases are depicted in Fig. 4.11 for the initial condition $r_0 = 2c$. For Case (a), $v_M = 2v_T$, so that $\chi = \pi/2$; for Case (b), two values of $v_C/v_T$ are examined, namely 1 and 2. For comparison, the parallel-navigation trajectory for $v_M = 2v_T$ is also shown in this figure.

## 4.5.2   Constant Projected Line

When M can make use of an external geometrical reference, yet another geometrical rule close to parallel navigation can be employed, according to which the *apparent trajectory* of T, i.e., relative to some fixed reference, is a straight line. This may be

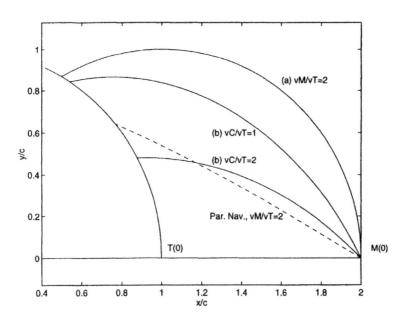

Figure 4.11: CAN Trajectories, Example in Sec. 4.5.1

the way baseball-playing fielders catch fly balls.[5] They seem to run along a curving path, adjusting their speed and direction so that they maintain a *linear optical trajectory* (LOT) for the ball relative to home plate and the background scenery [9].

When they follow this rule (unconsciously, of course), they keep a certain angle, called the *trajectory projection angle*, constant; this is the angle from the perspective of the fielder[2] that is formed by the ball, home plate[2], and a horizontal line emanating from home plate. Rather similarly, teleost fish and house flies "follow the motion of their target by maintaining an 'optical angle' that is a function of direction of movement" [ibid.].

For example, let us examine a case where the trajectory of the ball T is on the $x = 0$ plane in an $(x, y, z)$ Cartesian FOC, and the trajectory of the fielder M, on the $z = 0$ one; the $z$-axis is vertical (Fig. 4.12(a)). The coordinates of $T$ and $M$ are $(x_T, y_T, z_T)$ and $(x_M, y_M, z_M)$, respectively. At $t = 0$, $T$ is at the origin $O$ and $M$, at the point $(x_{M0}, 0, 0)$. The fixed reference, or background, is the plane $x = -L$, on which the apparent trajectory of T is seen by M. It is on this plane that M

---

[5]In baseball, a 'fly ball' is a ball batted in a high arc, which starts its flight at the 'home plate' and ends it (often) in the hands of the 'fielder'.

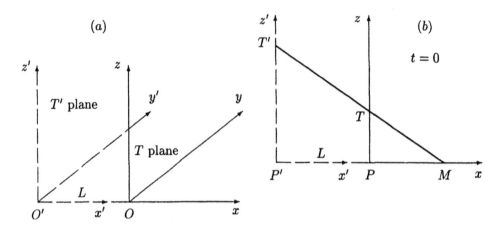

Figure 4.12: Geometry of Example in Sec. 4.5.2

maintains the LOT.

For the sake of simplicity we assume that the LOT starts at the point $O'(-L, 0, 0)$ and that M runs such that $y_M = y_T$. A *shifted* FOC centered at $O'$ is defined such that $x' = x + L$, $y' = y$, $z' = z$. $T'$ is the projection of $T$ from the perspective of $M$ on the background plane $x' = 0$; its coordinates are $(0, y'_T, z'_T)$. Thus the equation of the LOT is, by definition of the present geometrical rule, $x'_T = 0$, $z'_T = S y'_T$, $S$ being the constant slope of the LOT line.

Suppose now that the trajectory of the ball T is ballistic, effects such as wind or air resistance being neglected, and that T is thrown along the line $z = y$, $x = 0$, such that its initial velocities are $v_{y0} = v_{z0} = U$. Using elementary mechanics, the equation for M's trajectory can be shown to be

$$x_M = L \frac{1 - \frac{y_M}{Y}}{S - 1 + \frac{y_M}{Y}}, \qquad Y \triangleq \frac{2U^2}{g}. \tag{4.39}$$

Clearly, M catches T at the point $(0, Y, 0)$. Note that M changes both its speed and its direction as it runs to catch T; note also that even as T is falling (in the plane $x = 0$), M sees it *rising* on the background plane $x' = 0$.

The trajectories of $T$, $M$, and $T'$ are shown in Fig. 4.13 for $x_{M0} = L$ (and, consequently, $S = 2$).

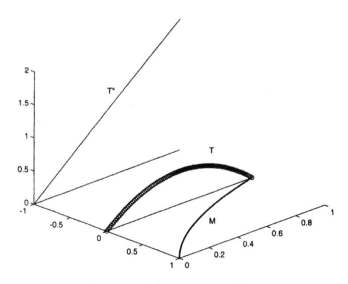

Figure 4.13: Trajectories of T', T, and M, Example in Sec. 4.5.2

## 4.6 References

[1] Locke, Arthur S., *Guidance*, a volume in the series *Principles of Guided Missiles Design* edited by Grayson Merrill, Van Nostrand, 1955, Ch. 12.

[2] Eades, James B. and George H Wyatt, "A study from Kinematics—The Problems of Intercept and Pursuit", *Goddard Space Flight Center*, 1969, NASA TMX 63527, N69-23785.

[3] Scharf, L. L., W. P. Harthill, and P. H. Moose, "A Comparison of Expected Flight Times for Intercept and Pure Pursuit Missiles", *IEEE Trans. Aero. Elec. Syst.*, Vol. AES-5, No. 4, 1969, pp. 672-673.

[4] Cockayne, E., "Plane Pursuit with Curvature Constraints", *SIAM J. Appl. Math.*, Vol. 15, No. 6, 1967, pp. 1511-1516.

[5] Newell, H. E., "Guided Missile Kinematics", *Naval Research Laboratory*, Report No. R-2538, May 1945.

[6] Rublein, G.T., "On Pursuit with Curvature Constraints", *SIAM J. Control*, Vol. 10, No. 1, 1972, pp.37-39.

[7] Gomes-Teixeira Francisco, "Traité des Courbes Spéciales Remarquables", Paris, Jacques Gabay, 1995 (A reimpression of the first edition, Coïmbre, Académie Royale des Sciences de Madrid, 1908-1915).

[8] Shneydor, N. A., "On Constant-Aspect Pursuit", *Proc. 30th Israel Conf. on Aviation and Astronautics*, 1989, pp. 215-217.

[9] McBeath, M. K., D. M. Shaffer, and M. K. Kaiser, "How baseball Outfielders Determine Where to Run to Catch Fly Balls", *Science*, vol. 268, 28 April 1995, pp. 569-573.

# Chapter 5

# Proportional Navigation

## 5.1 Background and Definitions

Proportional navigation (PN) is the guidance law which implements parallel navigation, the geometrical rule to which Chapter 4 was dedicated. Hence this is the law that makes $\mathbf{w}$ in the general case, or $\lambda$ in the planar one, tend to zero, provided $\mathbf{v}_T$ and $v_M$ are constant.

The flow diagram shown in Fig. 5.1, drawn for a planar engagement, illustrates the fact that proportional navigation, like all guidance laws, is an algorithm for a feedback control. In this diagram, $z_T$ and $z_M$ are small deviations of T and M, respectively, from a nominal LOS, such that the approximation $\lambda = (z_T - z_M)/r$ is valid. In the planar context, the PN guidance law is

$$\dot{\gamma}_{M_c} = N\dot{\lambda} , \tag{5.1}$$

where $\dot{\gamma}_{M_c}$ is the command for $\dot{\gamma}_M$ and $N$ is the *navigation constant*. Equivalently

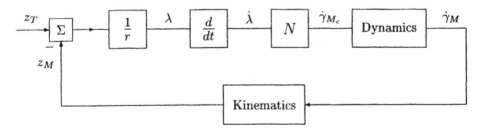

Figure 5.1: Proportional-navigation guidance loop

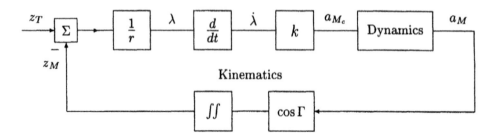

Figure 5.2: Proportional-navigation guidance loop in terms of $\mathbf{a}_M$

the law is often stated as

$$a_{M_c} = k\dot{\lambda} \; , \tag{5.2}$$

where $a_{M_c}$ is the command for lateral acceleration and $k$ is a constant representing
*gain*. Since $a_M = v_M \dot{\gamma}_M$,

$$k = N v_M \; .$$

Fig. 5.2 is a modification of Fig. 5.1, better suited for the definition of PN in terms
of lateral acceleration. The angle $\Gamma$ is the angle between $\mathbf{a}_M$ and the normal to $\mathbf{r}$;
its significance will become apparent in Sec. 5.4.

The reason for the name 'proportional navigation' is now perhaps clear; 'proportional' because of the way PN is defined, see (5.1) or (5.2), and 'navigation' because
of its origin in the parallel navigation geometrical rule. On the whole, not a very
good name, but accepted everywhere. It was already used by US navy personnel in
1945 [1, 2], and eventually had better success than other names used in the early
years of PN, such as 'navigational course', 'linear homing navigation', and 'partial
navigation course'.

For general, i.e., three-dimensional engagements, PN is defined by the equation

$$\mathbf{a}_{M_c} = N\mathbf{w} \times \mathbf{v}_M \; ; \tag{5.3}$$

another definition, not equivalent to (5.3), is

$$\mathbf{a}_{M_c} = k\mathbf{w} \times \mathbf{1}_r \; , \tag{5.4}$$

where $\mathbf{w}$ is the angular rate of $\mathbf{r}$ (see Appendix A.1). By (A.2) and (A.3) of Appendix
A.1, guidance law (5.4) can be expressed in the alternative form

$$\mathbf{a}_{M_c} = -\frac{k}{r}\mathbf{v}_{C\perp} \; , \tag{5.5}$$

where $\mathbf{v}_{C_\perp}$ is the component of the closing speed $\mathbf{v}_C$ across the LOS $\mathbf{r}$. Comparing this formulation of the guidance law with (4.21) may be enlightening.

The laws (5.3) and (5.4) are called *Pure PN* (PPN) and *True PN* (TPN), respectively, for rather arbitrary reasons.

## 5.2   A Little History

It has been stated that PN was known by German scientists at Peenemünde and that they had developed its basic equations and principles by the end of the second world war [3,4]. However, none of the better-known German sources makes this claim [5-9]. Certain developments were close to PN, especially those where mounting the seeker on a gyroscope was considered [10,11], but nowhere can a definition similar to (5.1) - (5.4) be found in the said well-documented references.

Thus, it is reasonable to assume that PN was invented by C. L. Yuan at the RCA Laboratories in the USA. His first report dealing with PN, dated December 1943, was declassified a few years later and published in 1948 in *Applied Physics* [12]. Other pioneering reports were published by Newell and by Spitz, in 1945 and 1946, respectively [13, 1], and, a few years later, by Bennet and Mathews [14] and Adler [15]. First treatment in textbooks is probably in the Series *Principles of Guided Missile Design*, in the volumes by Locke and by Jerger, respectively [16,17].

In the United States, the main motivation for the development of surface-to-air and air-to-air PN-guided missiles towards the end of the war was no doubt the "horribly effective Kamikaze attacks" against US ships. Development had not ended by the end of the war; the first successful intercept made by a missile against a (pilotless) aircraft was in December 1950, by a Raytheon-developed Lark missile [18].

In the mid 1950's, the first PN-guided missiles appeared in several arsenals, developed and produced in various locations. It is remarkable that some of them are still being produced and deployed forty years later, albeit with many modifications and using modern technologies.

We cannot conclude this section without admitting that predators may have preceded human beings in the use of proportional navigation: "[PN with $N > 1$] is probably the simplest way to solve the problem of navigation. A vertebrate has semicircular canals to measure rates of rotation [i.e., $\dot{\gamma}$] and also may difference the output from its accelerometers each side while lower creatures have comparable sensors. Thus the rotation changes measured inside the head [$\dot{\gamma}$] combined with directional information from the eyes, ears and other senses [i.e., $\lambda$] enable the hunter to intercept its prey as economically as possible" [19].

## 5.3    Kinematics of A Few Special Cases

There is no general closed-form solution to PN guidance equations, even when lag-free dynamics is assumed. We shall tackle the theory gradually, beginning with the simplest planar cases and relaxing assumptions as we proceed. In this chapter we shall assume that $v_M$ is constant and that M is ideal, i.e., its dynamics is linear and lag-free, such that $\gamma_M \equiv \gamma_{M_c}$ and $\mathbf{a}_M \equiv \mathbf{a}_{M_c}$.

### 5.3.1    Two Special Values of $N$

(a) $N = 1$

Since $\dot{\gamma}_M = \dot{\lambda}$ by (5.1), integration gives the equation

$$\gamma_M - \gamma_{M_0} = \lambda - \lambda_0 \, ,$$

which describes *pure pursuit*, deviated or not according to the initial conditions.

This property can be shown using another approach, as follows. In pure pursuit (which is instantaneously planar), one has $\mathbf{a}_M = \mathbf{w} \times \mathbf{v}_M$ provided $\dot{v}_M = 0$. Now this is precisely the expression for $\mathbf{a}_M$ in PN with $N = 1$ defined according to (5.3).

(b) $N \to \infty$

The gain being very large makes $\dot{\lambda}$ of the guidance loop (see Fig. 5.1) nearly zero (assuming that the loop remains stable). This corresponds to *parallel navigation*.

### 5.3.2    Stationary Target, Any $N$

An example for guidance towards a stationary target has already been given in connection with deviated pure pursuit (see Sec. 3.3.4(e)). It was noted there that this was not a triviality, since such applications do arise in practice. Indeed, some of the guided weapons specifically designed for use against stationary targets employ PN, in spite of the rather high cost of PN guidance units compared to, say, PP ones.

By (A.13) and (A.14) (see also Fig. 3.14), the kinematics of this case is given by the equations

$$\begin{cases} \dot{r} = -v_M \cos \delta \\ r\dot{\lambda} = -v_M \sin \delta \\ \dot{\gamma}_M = \dot{\lambda} + \dot{\delta} \, , \end{cases} \tag{5.6}$$

to which the PN guidance law $\dot{\gamma}_M = N\dot{\lambda}$ is added.

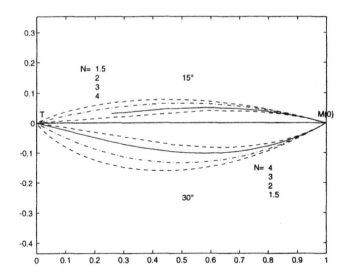

Figure 5.3: PN trajectories, stationary target

Solving (5.6) is quite straightforward. The solutions for $r$ and $a_M$ are

$$\frac{r}{r_0} = \left(\frac{\sin[\delta_0 + (N-1)\lambda]}{\sin \delta_0}\right)^{\frac{1}{N-1}}, \quad \lambda_0 \triangleq 0, \quad N > 1, \tag{5.7}$$

$$a_M = \frac{N v_M^2 \sin \delta_0}{r_0}\left(\frac{r}{r_0}\right)^{N-2}, \tag{5.8}$$

where $r_0$, $\delta_0$ and $\lambda_0$ are the initial conditions. The angle of arrival is found from (5.7) to be $\lambda_f = -\delta_0/(N-1)$. Several trajectories are shown in Fig 5.3 where T is at the origin, $r_0 = 1$ and $\delta_0 = -15°$ and $30°$.

Note an interesting property that results from (5.8). As $r \to 0$, the lateral acceleration $a_M \to 0$ or $\to \infty$ according to whether $N > 2$ or $N < 2$, respectively. If $N = 2$, then (5.7) describes a circle with radius $r_0/2 \sin \delta_0$, and $a_M$ is obviously constant. Its value is

$$a_M = \frac{2v_M^2}{r_0} \sin \delta_0 .$$

### 5.3.3 $N = 2$, Nonstationary, Nonmaneuvering Target

Suppose $\mathbf{r}$, $\mathbf{v}_T$ and $\mathbf{v}_M$ are coplanar and $\mathbf{v}_T$ and $v_M$ are constant. By these assumptions, $v_T$, $v_M = K v_T$, and $\gamma_T$ are constant, too. Hence the equations, by now probably familiar (see also Fig. 3.14),

$$\begin{cases} \dot{r} = v_T \cos\theta - v_M \cos\delta \\ r\dot{\lambda} = v_T \sin\theta - v_M \sin\delta \\ \gamma_M - \delta = \gamma_T - \theta = \lambda \,, \end{cases} \tag{5.9}$$

to which we add the guidance law

$$\dot{\gamma}_M = 2\dot{\lambda} \,. \tag{5.10}$$

Integrating this equation we obtain $\gamma_M = 2\lambda +$ a constant. Defining this constant as $-\alpha_0$ and choosing $\gamma_T \triangleq 0$ we obtain the relations

$$\begin{cases} \delta = \lambda - \alpha_0 \\ \theta = -\lambda \,, \end{cases} \tag{5.11}$$

which we substitute into (5.9). We finally get

$$\begin{cases} \dot{r} = v_T \cos\lambda - v_M \cos(\lambda - \alpha_0) \\ r\dot{\lambda} = -v_T \sin\lambda - v_M \sin(\lambda - \alpha_0) \,. \end{cases} \tag{5.12}$$

This couple of differential equations can be solved for $r(\lambda)$ and $a_M(\lambda)$. The solution, being rather cumbersome and not very interesting, is not given here; it can be found in Locke [17]. However, two interesting properties that the solution provides are as follows.

(i) If $\sin\delta_0 = (1/K)\sin\theta_0$, then collision course conditions exist and parallel navigation starts right at the beginning of the engagement. (The term 'collision course' has been defined in Sec. 4.2).

(ii) $a_M \to 0$ or $\to \infty$ as $r \to 0$ according to whether $K\cos\alpha_0 > -1$ or $< -1$, respectively. For $K\cos\alpha_0 = -1$, $a_M \to$ a finite constant.

## 5.4  Kinematics of PN, Approximative Approach

In this section we shall study various variants of planar PN, making the same assumptions as before but tackling more involved engagements. In order to do this we assume that M is near a collision course. By this new assumption, $\dot{r}$ is constant and the angle between $\mathbf{r}$ and $-\dot{\mathbf{r}}$ is small. Since this assumption will be used often in this section and in Sections 6 and 8, we shall henceforth simply refer to it as the *near collision course* (NCC) assumption.

## 5.4.1 True PN (TPN)

(a) This guidance law has been defined in (5.4) as $a_{M_c} = k\mathbf{w} \times \mathbf{1}_r$. In planar terms: $a_{M_c}$ is normal to the LOS $\mathbf{r}$, such that the angle $\Gamma$ in Fig. 5.2 is identically zero, and $a_{M_c}$ is given by the equation

$$a_{M_c} = k\dot{\lambda} . \tag{5.13}$$

By planar kinematics (see Sec. 1.3.2), the component of $\mathbf{a}_M$ perpendicular to $\mathbf{r}$ is

$$a_M = -(r\dot{\omega} + 2\omega\dot{r})$$

where $\omega = \dot{\lambda}$. (For a more general, three-dimensional vector treatment, see Appendix D.10(a)). Bearing in mind that $a_M \equiv a_{M_c}$ by the assumption of lag-free dynamics, we can combine this equation with the guidance law (5.13), obtaining the differential equation

$$r\dot{\omega} + 2\omega\dot{r} + k\omega = 0 . \tag{5.14}$$

As $\dot{r}$ is constant by assumption, we substitute $v_C = -\dot{r}$ and introduce *time-to-go* $\tau \triangleq t_f - t$ into (5.14), which, since $r = v_C\tau$ and $d/dt = -d/d\tau$, changes into

$$v_C\tau\frac{d\omega}{d\tau} + (2v_C - k)\omega = 0 . \tag{5.15}$$

The solution to this simple differential equation equation is

$$w(\tau) = \omega_0 \left(\frac{\tau}{t_f}\right)^{N'-2} , \tag{5.16}$$

where we have introduced the parameter $N'$ defined by

$$N' \triangleq \frac{k}{v_C} . \tag{5.17}$$

$N'$ is called the *effective navigation constant*.

Clearly, the nature of the solution depends on the value of $N'$. In order that $w \to 0$ as $\tau$ and $r \to 0$, *$N'$ must be greater than 2. This is a most important conclusion.*

It may have been noticed that similar results have been obtained in Sec. 5.3.2 for the case of a stationary target. We also note that, by (5.16), $w$ (and hence $a_M$) is constant for $N' = 2$.

(b) Based on the NCC assumption, an alternative definition of TPN can be formulated, as follows.

Suppose T is at the origin of a Cartesian frame of coordinates $(x, y)$ such that the $x$-axis is along the reference LOS and M is located at $(x, y)$. By the NCC

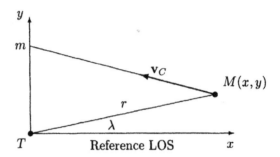

Figure 5.4: Zero-effort miss $m$

assumption, $\mathbf{v}_C$ is approximately parallel to the LOS and the LOS angle $\lambda$ is small, such that the approximations

$$x = r, \quad y = r\lambda = v_C \tau \lambda$$

are valid (see Fig. 5.4). It follows that

$$\dot{\lambda} = \frac{r\dot{y} - \dot{r}y}{r^2} = \frac{\dot{y}(t)\tau + y(t)}{v_C \tau^2} \; . \tag{5.18}$$

Now, by linear extrapolation, the *expected miss distance*, or *miss*, $m$ is easily seen to be

$$m(t, \tau) = y(t) + \dot{y}(t)\tau \; . \tag{5.19}$$

$m$ is also called the *zero-effort miss* since it represents the amount by which M misses T (at $\tau = 0$ by definition) if control has stopped at the moment $t = t_f - \tau$, provided T is not maneuvering.

Thus, finally, combining (5.13), (5.17), (5.18) and (5.19), the planar TPN guidance law can now be stated in the alternative form

$$a_{M_c}(t) = \frac{N' m(t, \tau)}{\tau^2} \; . \tag{5.20}$$

For a more general, vector representation of the constant-$\mathbf{v}_C$ kinematics of miss, see Appendix D.10(b).

*Note.* By (5.20), the acceleration command is proportional to the expected miss distance. This property is another motivation for PN, since it clearly shows that PN acts to reduce the miss to zero. However, it was found years after the invention of PN.

## 5.4.2 Use of Range-Rate in TPN

For guidance systems where $\dot{r}$ can be measured, e.g., systems that have radar seekers, a modification of the guidance law (5.13) can be used, namely

$$a_{M_c} = -k'\dot{r}\omega \qquad (5.21)$$

where $k'$ is a constant. The differential equation analogous to (5.14) is

$$r\dot{\omega} + 2\omega\dot{r} - k'\dot{r}\omega = 0 .$$

This equation is easily solved using the auxiliary variable $\xi \triangleq \log(r/r_0)$, such that $\dot{\xi} = \dot{r}/r$. The solution turns out to be similar to (5.16), namely

$$\omega = \omega_0 \left(\frac{r}{r_0}\right)^{k'-2} . \qquad (5.22)$$

Comparing (5.22) to (5.16) we naturally define

$$N' = k'$$

for this guidance law. The conclusions regarding permissible values for $N'$ remain the same.

Note that when studying this guidance law we did not assume that M was close to a collision course. This is certainly an advantage from the analytic point of view.

Some authors call this guidance law RTPN, the 'R' standing for "realistic". Since many real systems exist in which $\dot{r}$ is not available at all, we prefer the 'R' in RTPN to represent 'rate', so that we get *Rate-using True Proportional Navigation.*

## 5.4.3 Pure PN (PPN)

We recall that for both TPN and RTPN, Sections 5.4.1 and 5.4.2, respectively, the control acceleration was applied normal to $\mathbf{r}$. In the present case it is applied *normal to the velocity vector* $\mathbf{v}_M$, such that the angle $\Gamma$ in Fig. 5.2 equals $\delta$, which is the angle by which $\mathbf{v}_M$ leads $\mathbf{r}$ (see Fig. 5.5). This of course means that the component of $\mathbf{a}_M$ normal to $\mathbf{r}$ is $a_M \cos\delta$, i.e.,

$$\ddot{z}_M = a_M \cos\delta = v_M \dot{\gamma}_M \cos\delta ,$$

and that the differential equation for the present guidance law is

$$r\dot{\omega} + 2\omega\dot{r} + k\omega \cos\delta = 0$$

rather than (5.14). By the definition of PN, $\dot{\gamma}_M = N\dot{\lambda}$, and by the structure of the guidance loop, Fig. 5.2, $a_M = k\dot{\lambda}$. Therefore in PPN one has

$$N' = \frac{k \cos\delta}{v_C} = N\frac{v_M \cos\delta}{v_C} \qquad (5.23)$$

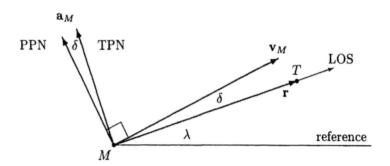

Figure 5.5: Geometries of TPN and PPN

rather than $N' = k/v_C = Nv_M/v_C$ shown in (5.17) for TPN.

In fact, we can generalize the definition of $N'$ by stating it as

$$N' \triangleq \frac{k \cos \Gamma}{v_C} = N \frac{v_M \cos \Gamma}{v_C} , \qquad (5.24)$$

since, by the respective definitions, $\Gamma = 0$ for TPN and $\Gamma = \delta$ for PPN.

We shall later consider the unavoidable question 'which of the two is better', TPN or PPN.

For an illustration, several trajectories that result from planar PN are shown in Fig. 5.6 for a constant-$v_T$ target. Note that in Fig. 5.6(a), the higher the value of N, the closer the trajectory is to the $N = \infty$ line, which is in fact a parallel-navigation straight-line trajectory. In Fig. 5.6(b), only the trajectory of M for $N = 4$ is shown, along with the target trajectory and lines-of-sight at equal time-intervals. Note that, as M approaches T, the trajectory becomes a straight line and the lines-of-sight become parallel to each other; this indicates that parallel navigation is being achieved.

The guidance cases depicted in Fig. 5.6(a) are presented again in Fig. 5.7(a) in the *TT frame of coordinates*. Trajectories for $K = 2$ and $N = 2, 3, 4$, and 6 are shown, as well as isochrones for $t = 0(0.1)0.6$, the unit of time being $r_{rbm}/v_T$ ('rbm' for 'right beam'). For comparison, the trajectory of pure pursuit for $K = 2$ is also shown. Asterisks (*) denote the points where the isochrones intersect the $\psi = 90^\circ$ ray, which is also the trajectory for $N = \infty$ (parallel navigation).

In Fig. 5.7(b), trajectories are shown for $K = 2, 3$, and 4, N being kept constant ($= 2$). Rather than isochrones, time marks are given on the trajectories, $r_{rbm}/v_T$ from each other.

We have seen that, for non-maneuvering targets, PN trajectories approach parallel-navigation straight line trajectories as M approaches T. In fact, if the initial conditions of a PN engagement are such that $\dot{\lambda} = 0$, then the PN trajectory is a

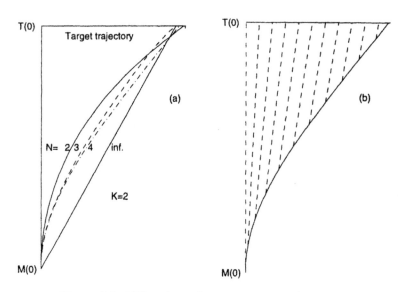

Figure 5.6: PN trajectories, nonmaneuvering target

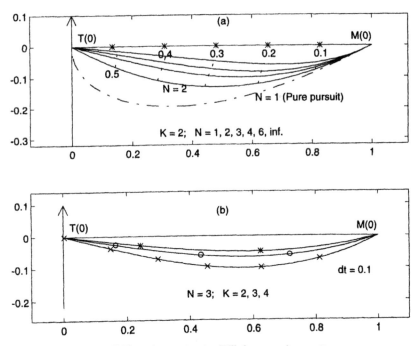

Figure 5.7: PN trajectories in TT frame of coordinates

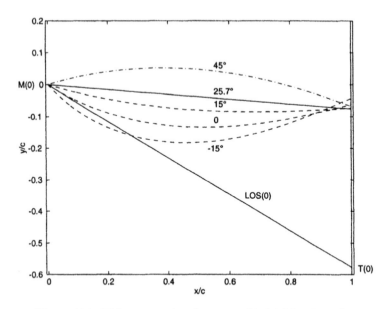

Figure 5.8: PN trajectories for several initial lead angles

straight line. In order that $\dot{\lambda}$ be zero, the lead angle $\delta$ must satisfy (4.8), namely $\sin \delta = (1/K) \sin \theta \triangleq \delta_{id}$ ('$id$' for 'ideal'). If $\delta < \delta_{id}$, then the lead angle is said to be less than ideal; if, moreover, $\delta < 0$, then M is said to start the engagement having a *lag* angle; finally, if $\delta > \delta_{id}$, this is an over-lead situation.

The advantages of starting a PN guidance engagement at a lead angle close to $\delta_{id}$ are illustrated in Fig. 5.8, where $K = 2$, $N = 3$, and T travels on the line $x_T(t) = c$, $y_T(t) = -c/\sqrt{3} + v_T t$; $x_M(0) = y_M(0) = 0$, such that $\lambda(0) = -30°$ and $\theta(0) = 120°$.

Since $K = 2$, the 'ideal' lead-angle is $\delta_{id} = \arcsin[(1/2) \sin 120°] = 25.7°$. For this value of $\delta$, the trajectory is the (full) straight line. Three dashed lines are shown for less-than-ideal lead-angles, namely $15°$, 0, and $-15°$, the latter representing of course a lag-angle condition. The dash-dot line is the trajectory for $\delta = 45°$, clearly an over-lead.

## 5.4.4   Some Results

Several interesting results can be derived from the equations developed in this section. We shall present three, in all of which $N' > 2$ is taken for granted [18].

(a) If the guidance system has a heading, or aiming, error $\gamma_0$ at $t = 0$, then M maneuvers according to the equation

$$a_M(\tau) = \frac{v_M \gamma_0 N'}{t_f} \left(\frac{\tau}{t_f}\right)^{N'-2}, \qquad (5.25)$$

which is, of course, similar to (5.16).

*Note.* It is not a mere curiosity that the minimum value of the *control effort* defined by the integral

$$\int_0^{t_f} a_M^2(\tau) d\tau ,$$

$a_M(\tau)$ being given by (5.25), is obtained for $N' = 3$. The optimality of this value will be dwelt upon in Chapter 8.

(b) If the system has an acceleration bias $a_b$ perpendicular to the LOS (which could, for example, be the gravity acceleration), M has to maneuver according to the equation

$$a_M(\tau) = a_b \left[ \frac{N'}{N'-2} \left(\frac{\tau}{t_f}\right)^{N'-2} - \frac{2}{N'-2} \right]. \qquad (5.26)$$

Note the $(N' - 2)$ in the denominators; also note that $|a_M/a_b| \to 2/(N' - 2)$ as $\tau \to 0$.

(c) Finally, if T maneuvers at a constant lateral acceleration $a_T$, $v_T$ remaining constant, then M is required to maneuver, too. The acceleration ratio is

$$\frac{a_M(\tau)}{a_T} = \frac{N'}{N'-2} \left[ 1 - \left(\frac{\tau}{t_f}\right)^{N'-2} \right], \qquad (5.27)$$

where $a_T$ and $a_M$ denote here the components of $\mathbf{a}_T$ and $\mathbf{a}_M$ across the LOS, respectively. Note that towards intercept (at $\tau = 0$, or $t = t_f$), $a_M/a_T$ approaches the value

$$\left(\frac{a_M}{a_T}\right)_f = \frac{N'}{N'-2} \qquad (5.28)$$

(see also Fig. 5.9). This equation is the reason why PN-guided missiles are often said to require lateral acceleration capabilities exceeding 3 times the target capabilities; this number results from (5.28) for $N' = 3$.

## 5.5 Kinematics of PN, Exact Approach

In this section we abandon the NCC approximating assumptions, on which the theory presented in Sects. 5.4.1 and 5.4.3 was based. We still assume that $\mathbf{r}$, $\mathbf{v}_T$

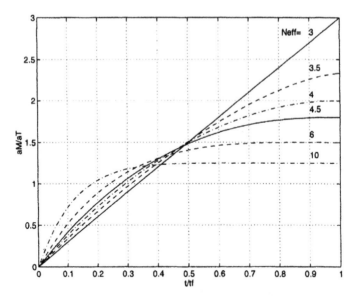

Figure 5.9: Lateral accelerations $a_M$ required when T is maneuvering

and $\mathbf{v}_M$ are coplanar, that $v_T$ and $v_M$ are constant, and that the dynamics of M is ideal such that $a_M \equiv a_{M_c}$. The kinematics is described by (5.9), which we reproduce here as (5.29):

$$\begin{cases} \dot{r} = v_T \cos\theta - v_M \cos\delta \\ r\dot{\lambda} = v_T \sin\theta - v_M \sin\delta \\ \gamma_M - \delta = \gamma_T - \theta = \lambda \ . \end{cases} \tag{5.29}$$

We will review several analytic approaches, in some of which $\mathbf{v}_T$ will not necessarily be constant. In none of them shall we have simple, closed-form results of the kind obtained in the previous section. This is quite natural, of course, since when less assumptions are made, results are bound to be of a more general and qualitative nature.

## 5.5.1   TPN

The first general closed-form solution to the problem of TPN guidance was published by Guelman in 1976 [20], where conditions were obtained that ensured intercept, or monotone-decreasing $\lambda$, as functions of the initial state of the engagement. The initial state was given in the $(v_{C_\parallel}, v_{C_\perp})$ plane, where $v_{C_\parallel}$ and $v_{C_\perp}$ are the radial and tangential components of bf $\mathbf{v}_C$, respectively (see Sec. 1.3.2 regarding these terms). In recent years, several papers have been published which, along a similar approach, extend the early results and, sometimes, present modifications of the basic TPN law. We shall describe some of them now.

(a) Cochran et al. have obtained analytical solutions for TPN guidance problems where $N'$ is 3 or 4, the guidance law used being RTPN (see Sec. 5.4.2 for the definition of RTPN) [21]. In both the $N' = 3$ and the $N' = 4$ cases, Jacobian elliptic integrals are involved in the expressions for trajectories.

(b) Capture regions for RTPN have been calculated by Dhar and Ghose and, surprisingly enough, found to be smaller (i.e., worse) than those obtained for TPN [22]. In fact "it was observed that RTPN showed significant degradation in performance compared with Guelman's TPN".

(c) The solutions presented by Yuan and Chern are very compact [23]. The guidance law analysed is RTPN, and the main results presented are as follows. Let $C(\theta)$ and $S(\theta)$ denote $\cosh(\sqrt{N'-1}\,\theta)$ and $\sinh(\sqrt{N'-1}\,\theta)$, respectively. Then

$$
\begin{cases}
\dot{\theta} = \dot{\theta}_0 [C(\theta) + \frac{1}{B}S(\theta)]^{(N'-2)/(N'-1)} \\[2mm]
r\dot{\theta} = r_0\dot{\theta}_0 [C(\theta) + \frac{1}{B}S(\theta)] \\[2mm]
\dot{r} = \dot{r}_0 [C(\theta) + BS(\theta)] \,,
\end{cases}
\tag{5.30}
$$

where

$$
B \triangleq \frac{r_0\dot{\theta}_0}{\sqrt{N'-1}\,\dot{r}_0} \,.
$$

In the same paper where these results are presented, the problem of a maneuvering target is also tackled. However, the analysis is based on the rather odd assumption that the magnitude of the target acceleration is regulated proportional to the closing speed. The only justification given for this assumption is that it simplifies the formulation.

(d) In a paper on TPN and RTPN with maneuvering targets, yet another unusual assumption is made regarding $a_T$, namely that it is inversely proportional to the component of $v_C$ normal to the LOS [24]. This is not illogical, though, since "T pulls a high acceleration when M is on a collision course [high $\dot{r}$] and a negligible acceleration when the LOS rate is already high".

(e) A generalization of TPN has been proposed by Yang et al. where the commanded acceleration $a_{M_c}$ forms a constant angle $\Gamma$ with the normal to the LOS (see Fig. 5.2) [25]. $\Gamma$ may depend on the initial values of $v_{C_\parallel}$ and $v_{C_\perp}$. The authors of this law call it *generalized proportional navigation* (GPN). When $\Gamma = 0$, this law clearly reduces to TPN. It is claimed that, under certain conditions, GPN has a larger capture area and a shorter interception time than TPN. More details on

GPN (later called generalized *true* proportional navigation, GTPN) can be found in [26-28].

(**f**) A combination of (c) and (e) has been proposed by the authors of [28] as a *rendezvous* guidance law [29]. (In rendezvous guidance, the relative velocity between M and T must be driven to zero at the intercept). According to this law, $\mathbf{a}_{M_c}$ has two components, $a_{M_{c\perp}}$ and $a_{M_{c\parallel}}$, across and along the LOS, repectively, such that $a_{M_{c\perp}} = -N'\dot{r}\dot{\theta}$ and $a_{M_{c\parallel}}$ depends in a certain way on $N'$ and the initial conditions $r_0$, $\dot{r}_0$, and $\dot{\theta}_0$. It turns out that, with this algorithm, the relative trajectory is the logarithmic spiral $\theta = \theta_0 + (r_0\dot{\theta}_0/\dot{r}_0)\log(r/r_0)$.

## 5.5.2  PPN

(**a**) Yoriaki Baba et al. generalize PN for the case where both T and M have constant *axial* accelerations, $A_T$ and $A_M$, respectively [30]. In the new, generalized form, the acceleration command is

$$\mathbf{a}_{M_c} = \frac{N}{r^2}[(\mathbf{v}_M - k''\mathbf{v}_T) \times \mathbf{r}] \times \mathbf{v}_M \ , \tag{5.31}$$

where $k''$ is given by the (nonconstant) ratio

$$k'' = \frac{1 + \frac{A_T \tau}{2v_T}}{1 + \frac{A_M \tau}{2v_M}} \ . \tag{5.32}$$

When both axial accelerations are zero, $k'' = 1$ and (5.31) becomes equivalent to the PPN guidance law (5.3). The authors call this law "the true guidance law". While this law may be the true one, it is certainly not easy to implement; therefore the authors also propose a simplified version.

(**b**) The approximative linearized analysis of PPN which was presented in Sec. 5.4.3 dates from the 1950's and is therefore often referred to as the *classical* solution of PN. In a modern approach proposed by Shukla and Mahapatra, a quasi-linearized closed-form solution is obtained [31]. This solution is further reduced to a linear form through small-angle approximations which are less restrictive than those of the classical solution. It is shown that the more sophisticated linearization does indeed provide solutions which are closer to the true ones (found by numerical simulations) than the classical solutions. This quasi-linearization method has also been employed for the study of PPN with maneuvering targets [32].

(**c**) Two years after Shukla and Mahapatra had published their first paper, an *exact* closed-form solution to the PPN problem was found by Becker [33]. In his paper it is shown that there exists at least a partial closed-form solution $r(\theta)$ for all navigation constants $N \geq 2$. The solution is very cumbersome, however, its structure being given by a convergent infinite product. Certain simplification is obtained if $N$ is assumed to be a rational number.

(d) *Modified PPN* has been defined in the context of maneuvering targets by the guidance law

$$a_{M_c} = (1 + \frac{N}{\cos \delta})v_M \dot{\lambda} + \frac{a_T}{\cos \delta} \, , \qquad (5.33)$$

where $a_T$ is the (estimated) acceleration of T normal to the LOS [34]. Since $\delta$ is not always easily available as a signal in a particular realization, an approximation to (5.33) is suggested where $\psi_S$ is substituted for $\delta$, the former being the angle between M's axis and the LOS; in many instrumentation realizations this angle is measured by the seeker on board M. (Note that in the planar case, the difference between $\delta$ and $\psi_S$ is the angle of attack $\alpha$; see e.g. Fig. 3.20). It seems that this modification of PN requires lower values of $N$ than required by PPN, and that $a_M$ values tend to be lower, too.

The second term of (5.33) brings us to the topic of *Augmented PN*, which will be dealt with in Sec. 7.2.1.

## 5.5.3 TPN vs. PPN

The preceeding two subsections have been respectively devoted to TPN and to PPN; similarly, Sections 5.4.1 and 5.4.2 dealt with TPN, and 5.4.3, with PPN. A comparison between the two guidance laws is now in order. It will be based on some of the references mentioned above and on a paper by Shukla and Mahapatra [35].

(i) From the point of view of ease of analysis, TPN is quite superior; we have mentioned a few exact solutions available for it. PPN presents tremendous analytical difficulties. Approximative methods of analysis are available for both, of course.

(ii) Capture zones provided by TPN are smaller than those provided by PPN. It has been shown that, whereas capture of the target can be assured for the entire plane of the initial values of $v_{C_\parallel}$ and $v_{C_\perp}$ in PPN (except for a well-defined particular case), in TPN capture is restricted to a certain "circle of capture" [20]. However, Ghose claims that after a minor modification of GTPN (see Sec. 5.5.1(e)) the latter "becomes almost comparable to PPN law insofar as the capture region is concerned" [36].

(iii) PPN is more robust. It is not very sensitive to the "finer aspects of the initial conditions" [35].

(iv) The *control effort* (usually measured by the time-integral of $|a_M|$; also called *velocity increment*) required by TPN is somewhat higher, especially for low values of $N$.

## 5.6    PPN and TPN in 3-D Vector Terms

### 5.6.1    Definitions and Some Properties

(a) In Sec. 5.1, PPN and TPN have been defined by the guidance laws (5.3) and
(5.4), respectively, namely

$$\mathbf{a}_{M_c} = N\mathbf{w} \times \mathbf{v}_M \tag{5.34}$$

and

$$\mathbf{a}_{M_c} = k\mathbf{w} \times \mathbf{1}_r , \tag{5.35}$$

respectively, where

$$\mathbf{w} = \frac{\mathbf{r} \times \dot{\mathbf{r}}}{r^2} = \frac{\mathbf{r} \times (\mathbf{v}_T - \mathbf{v}_M)}{r^2} = \frac{\mathbf{r} \times (-\mathbf{v}_C)}{r^2} \tag{5.36}$$

by (A.3).

(b) Suppose $\mathbf{v}_C$ is resolved into two mutually orthogonal components, $\mathbf{v}_{C_\perp}$ and
$\mathbf{v}_{C_\parallel}$, across the LOS $\mathbf{r}$ and along it, respectively. It then follows from (5.34), (5.35)
and (5.36) that $a_{M_c}$ in both PPN and TPN is proportional to $v_{C_\perp}$. It is of interest
to compare this property with (4.34) and (4.35) and the 'nonfeedback law' described
in Sec. 4.4.2.

(c) Furthermore, by (5.35) and (5.36), TPN guidance law can be expressed as

$$\mathbf{a}_{M_c} = k\frac{\mathbf{r} \times (-\mathbf{v}_C)}{r^2} \times \mathbf{1}_r . \tag{5.37}$$

Hence, by (A.2) of Appendix A.1, $\mathbf{a}_{M_c}$ in TPN is directed along $-\mathbf{v}_{C_\perp}$:

$$\mathbf{a}_{M_c} = -\frac{k}{r}\mathbf{v}_{C_\perp} . \tag{5.38}$$

(d) Yet another expression for $\mathbf{a}_{M_c}$ in TPN is in terms of the (vector) miss
distance $\mathbf{m}$, namely

$$\mathbf{a}_{M_c} = k\frac{\mathbf{v}_C \times \mathbf{m}}{r^2} \times \mathbf{1}_r . \tag{5.39}$$

In spite of first-glance impression, there is no contradiction between (5.37) and
(5.39). For details, see Appendix D.10.

(e) We now turn our attention to $\mathbf{a}_{M_c}$ in PPN. We will show that this command
acceleration is coplanar with $\mathbf{r}$ and $\mathbf{v}_C$. (The trivial case where $\mathbf{r}$ and $\mathbf{v}_C$ are colinear
and, consequently, do not define a plane, is of no interest, since then one has parallel
navigation).
By (5.34),

$$\mathbf{a}_{M_c} \bullet (\mathbf{r} \times \mathbf{v}_C) = N(\mathbf{w} \times \mathbf{v}_M) \bullet (\mathbf{r} \times \mathbf{v}_C)$$

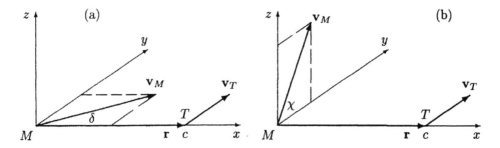

Figure 5.10: Illustration for Example in Sec. 5.6.2

$$= N(\mathbf{w} \times \mathbf{v}_M) \bullet (-r^2 \mathbf{w}) \quad \text{by (5.36)}.$$

The value of this product is clearly 0 by the properties of triple scalar products (see Appendix A.3), hence coplanarity has been shown.

The plane defined by $\mathbf{r}$ and $\mathbf{v}_C$ is often referred to as the *engagement plane*; thus, $\mathbf{a}_{M_{c,PPN}}$ is in the engaement plane.

**(f)** Suppose $\mathbf{v}_M$ forms the angle $\theta_{v_M}$ with the engagement plane. It follows from (5.34) and (5.35) that the ratio $a_{M_{c,PPN}}/a_{M_{c,TPN}}$ equals $\cos\theta_{v_M}$ if $k = N v_M$.

**(g)** Finally, we wish to recall that a guidance situation is said to be instantaneously planar if the vectors $\mathbf{v}_M$, $\mathbf{v}_T$, and $\mathbf{r}$ are coplanar. A criterion for this coplanarity is that the scalar triple product $\mathbf{v}_T \bullet (\mathbf{r} \times \mathbf{v}_M)$ should equal zero (for details, see Appendix A.3). If T is not maneuvering, then both PPN and TPN engagements tend with time to parallel navigation, which is obviously planar.

We will now illustrate the difference between PPN and TPN by a simple example.

## 5.6.2 An Example

**(a)** Let $\mathbf{r} = [c\ 0\ 0]^T$, $\mathbf{v}_T = [0\ U\ 0]^T$, and $v_M = KU$. $c$, $U$, and $K$ are constant. Then, if $\mathbf{v}_M = KU[\cos\delta\ \sin\delta\ 0]^T$ (see Fig. 5.10(a)), this is clearly a planar case, the (fixed) plane being $z = 0$. Note that this is the 'engagement plane' defined above and that $\theta_{v_M} = 0$. By (5.36), since $\mathbf{v}_T - \mathbf{v}_M = [-KU\cos\delta\ \ U-KU\sin\delta\ \ 0]^T$,

$$\mathbf{w} = \frac{1}{c^2} \begin{vmatrix} \mathbf{1}_x & c & -KU\cos\delta \\ \mathbf{1}_y & 0 & U - KU\sin\delta \\ \mathbf{1}_z & 0 & 0 \end{vmatrix} = \frac{U}{c} \begin{bmatrix} 0 \\ 0 \\ 1 - K\sin\delta \end{bmatrix}. \tag{5.40}$$

Hence, by (5.34) and (5.35), the lateral accelerations required by PPN and TPN are, respectively,

$$
\mathbf{a}_{M_c,PPN} = N \begin{vmatrix} \mathbf{1}_x & 0 & KU\cos\delta \\ \mathbf{1}_y & 0 & KU\sin\delta \\ \mathbf{1}_z & (U/c)(1 - K\sin\delta) & 0 \end{vmatrix} =
$$

$$
= \frac{NKU^2}{c} \begin{bmatrix} -(1 - K\sin\delta)\sin\delta \\ (1 - K\sin\delta)\cos\delta \\ 0 \end{bmatrix}
$$

and

$$
\mathbf{a}_{M_c,TPN} = k \begin{vmatrix} \mathbf{1}_x & 0 & 1 \\ \mathbf{1}_y & 0 & 0 \\ \mathbf{1}_z & (U/c)(1 - K\sin\delta) & 0 \end{vmatrix} = \frac{kU}{c} \begin{bmatrix} 0 \\ 1 - K\sin\delta \\ 0 \end{bmatrix}.
$$

It is easily deduced that both acceleration-command vectors are in the $z = 0$ plane, forming the angle $\delta$ between one another. Note that if $K\sin\delta = 1$, then $\|\mathbf{w}\| = 0$, i.e., CC conditions prevail, PN becomes identical with parallel navigation, and both acceleration commands are 0.

**(b)** Now let $\mathbf{v}_M$ be $KU[0 \ \cos\chi \ \sin\chi]^T$, $\mathbf{r}$ and $\mathbf{v}_T$ remaining unchanged (Fig. 5.10(b)). Clearly this is a nonplanar engagement. The LOS rate is now

$$
\mathbf{w} = \frac{1}{c^2} \begin{vmatrix} \mathbf{1}_x & c & 0 \\ \mathbf{1}_y & 0 & U - KU\cos\chi \\ \mathbf{1}_z & 0 & -KU\sin\chi \end{vmatrix} = \frac{U}{c} \begin{bmatrix} 0 \\ K\sin\chi \\ 1 - K\cos\chi \end{bmatrix},
$$

hence the respective acceleration commands

$$
\mathbf{a}_{M_c,PPN} = \frac{NKU^2}{c} \begin{vmatrix} \mathbf{1}_x & 0 & 0 \\ \mathbf{1}_y & K\sin\chi & \cos\chi \\ \mathbf{1}_z & 1 - K\cos\chi & \sin\chi \end{vmatrix} = \frac{NKU^2}{c} \begin{bmatrix} K - \cos\chi \\ 0 \\ 0 \end{bmatrix}
$$

and

$$
\mathbf{a}_{M_c,TPN} = \frac{kU}{c} \begin{vmatrix} \mathbf{1}_x & 0 & 1 \\ \mathbf{1}_y & K\sin\chi & 0 \\ \mathbf{1}_z & 1 - K\cos\chi & 0 \end{vmatrix} = \frac{kU}{c} \begin{bmatrix} 0 \\ 1 - K\cos\chi \\ -K\sin\chi \end{bmatrix}.
$$

This particular TPN case is singular in the sense that both the acceleration command $\mathbf{a}_{M_c,TPN}$ and the velocity $\mathbf{v}_M$ have no $x$-components, such that M cannot as it were leave the $x = 0$ plane.

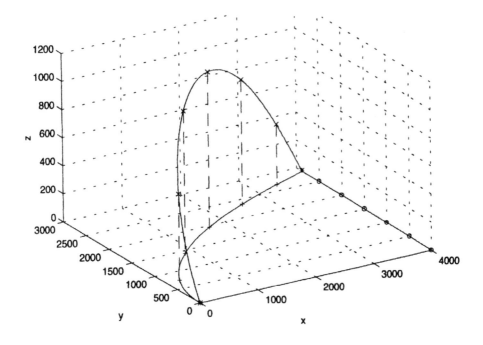

Figure 5.11: PPN Trajectories, Example in 5.6.2(c)

(c) A 3-D illustration of the PPN case is shown in Fig. 5.11 for a lag-free M, i.e., such that $\mathbf{a}_M \equiv \mathbf{a}_{M_{c,PPN}}$. The engagement starts with the initial conditions (see also Fig. 5.10(b)) $r = 4000\,m$, $U = 200\,m/sec$, $K = 2$, $\chi = 40°$, and $N = 3$. M's trajectory is marked with x's and its projection on the $z = 0$ with $+$ signs. T's trajectory is marked by o's, all marks shown at 2.5 $sec$ intervals.

The fact that this guidance process tends to parallel navigation is evidenced by the fact that all three components of $\mathbf{a}_M$ tend to 0 with time - see Fig. 5.12.

## 5.7 Other Laws that Implement Parallel Navigation

The guidance laws that will be reviwed in this section differ significantly from TPN, PPN, GPN and their variants; however, they have an important property in common with those laws, namely, that they are algorithms for implementing parallel navigation.

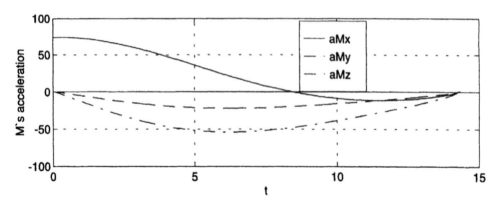

Figure 5.12: M's acceleration, Example in 5.6.2(c)

### 5.7.1   Ideal PN

Here we meet a guidance law which is not an extension of either TPN or PPN [37]. It is defined by the equation

$$\mathbf{a}_{M_c} = N\mathbf{w} \times \mathbf{v}_C \tag{5.41}$$

and is seen to be similar to (5.3) except that $\mathbf{v}_C$ has been substituted for $\mathbf{v}_M$. It can also be expressed as

$$\mathbf{a}_{M_c} = -\frac{N}{r^2}(\mathbf{r} \times \dot{\mathbf{r}}) \times \dot{\mathbf{r}} \ . \tag{5.42}$$

In planar terms this new law simply means that $\mathbf{a}_M$ should be normal to the vector $\mathbf{v}_C = \mathbf{v}_M - \mathbf{v}_T$ and that $a_{M_c} \propto v_C \dot{\lambda}$. In the reference cited above, closed-form solutions are given for both maneuvering and nonmaneuvering targets. The solutions are particularly compact for the latter case, as follows.

Let $\phi$ be the angle between $\mathbf{r}$ and $\dot{\mathbf{r}}$. Then

$$\dot{r} = v_C \cos\phi, \quad r\dot{\lambda} = v_C \sin\phi \ , \tag{5.43}$$

and $\phi$ is found to vary with $r$ according to the equation

$$\sin\phi = \left(\frac{r}{r_0}\right)^{N-1} \sin\phi_0 \ . \tag{5.44}$$

The required lateral acceleration is

$$a_M = Nv_{C_0}\dot{\lambda}_0 \left(\frac{r}{r_0}\right)^{N-2} \ . \tag{5.45}$$

Although Ideal PN intercepts targets faster than TPN does, the energy required is higher. Compared to results obtained by GPN (see Sec. 5.5.1(e) above), Ideal

PN has been found to achieve larger capture zones; it can in fact be considered as a near-optimal solution of GPN. However, the authors of this guidance law ignore the problem of how it can be implemented in practice.

## 5.7.2   Prediction Guidance Law

This guidance law has been proposed in 1985 by Kim et al. [38]. The main idea of this law is continuously to predict a straight-line collision course (CC) and to turn the heading of M into this direction as rapidly as possible. It is therefore, in a way, closer to parallel navigation than PN. As originally stated, the law is simply $a_{M_c} = K_R \Delta \delta$ where $\Delta \delta$ is the angle between the collision course and the current heading of M; in other words, $\Delta \delta$ is the difference between the lead-angle required for CC conditions and the current lead-angle $\delta$. It is therefore given by the equation

$$\Delta \delta = \delta - \arcsin(\sin \delta + r \dot{\lambda}/v_M) . \tag{5.46}$$

By another approach, the guidance law is

$$a_{M_c} = \frac{k}{K}(\sqrt{K^2 - \sin^2 \theta} - \cos \theta)\dot{\lambda} , \quad \mathbf{a}_{M_c} \perp \mathbf{r} , \tag{5.47}$$

where $k$ is a constant [26, 39]. It is based on the fact that, by the geometry of the collision triangle discussed in Sec. 4.2.1, the desired CC direction $\mathbf{1}_{M_d}$ is given by vector sum

$$K\mathbf{1}_{M_d} = \mathbf{1}_T + (\sqrt{K^2 - \sin^2 \theta} - \cos \theta)\mathbf{1}_r$$

where $\mathbf{1}_{M_d}$ and $\mathbf{1}_T$ are $\mathbf{v}_{M_d}/v_M$ and $\mathbf{v}_T/v_T$, respectively. It has been shown that, for sufficiently high $K$, guidance law (5.47) steers M rapidly into the desired collision course. Note that this law can be regarded as a variable-gain TPN law; as $K \to \infty$, the law approaches regular TPN.

It has also been claimed that, with a proper choice of $K$, this law is superior to TPN, at least with respect to capture zones and lateral accelerations. However, it certainly is not simple to implement since $K$ and $\theta$, which are parameters in the expression for the guidance law, are not immediately available in instrumentation realizations of most guidance systems.

## 5.7.3   Schoen's Laws

E. T. Schoen has proposed two guidance laws that implement parallel navigation and, unlike PN, have compact closed-form solutions [40].

To obtain the first of these laws, we substitute $\gamma_M - \lambda$ for $\delta$, and $-\lambda$ for $\theta$ in (5.29). The second substitution results from the assumption that T does not maneuver, hence $\gamma_T$ may be defined to be zero. We get the equations

$$\dot{r} = v_T \cos \lambda - v_M \cos(\gamma_M - \lambda) \tag{5.48}$$

$$r\dot{\lambda} = -v_T \sin\lambda - v_M \sin(\gamma_M - \lambda) \; . \qquad (5.49)$$

Differentiating (5.49) and using (5.48), the differential equation

$$\ddot{\lambda} r + \dot{\lambda}\dot{r} = -\dot{\lambda}\dot{r} - v_M \dot{\gamma}_M \cos(\gamma_M - \lambda)$$

is obtained which, multiplied by $r$ and rearranged, becomes

$$r^2 \ddot{\lambda} + 2\dot{\lambda}\dot{r}r = -r v_M \dot{\gamma}_M \cos(\gamma_M - \lambda) \; . \qquad (5.50)$$

The left-hand side of (5.50) is recognized to equal $\frac{d}{dt}(r^2\dot{\lambda})$. Now the guidance law

$$\dot{\gamma}_{M_c} = \frac{k_1 r \dot{\lambda}}{\cos\delta} \qquad (5.51)$$

is defined. Since $\dot{\gamma}_M = \dot{\gamma}_{M_c}$ by the assumption that M has lag-free dynamics, (5.50) can be solved using (5.51), and we finally get the expression

$$r^2 \dot{\lambda} = r_0^2 \dot{\lambda}_0 \epsilon^{-k_1 v_M t} \; . \qquad (5.52)$$

Note that the law (5.51) is in fact a *weighted* PN *law*. We used similar weighting when we examined a guidance law for pure pursuit in Sec. 3.4.2, albeit with another function of $r$.

The second law,

$$\dot{\gamma}_{M_c} = k_2 \frac{|\dot{\lambda}|^{1/2} sgn(\dot{\lambda})}{\cos\delta} \; , \qquad (5.53)$$

also leads to a closed-form solution. It has the additional interest that $|\dot{\lambda}|$ has the exponent $1/2$ rather than the usual 1.

It has not been claimed that either of the laws has practical importance.

## 5.8   References

[1] Spitz, Hillel, "Partial Navigation Courses for a Guided Missile Attacking a Constant Velocity Target", *Naval Research Laboratory*, Report No. R-2790, March 1946.

[2] Lancaster, Lt. Cdr. O. E. et al., "Proportional Navigation, Its Use in Pilotless Aircraft", *BUAER*, ADR Report No. T117, Nov. 1945 (Reference B in [1]).

[3] Nesline, F. W. and P. Zarchan, "A New Look at Classical vs Modern Homing Missile Guidance", *J. Guidance*, Vol. 4, No. 1, 1981, pp. 78-85.

[4] Fossier, M. W., "Tactical Missile Guidance at Raytheon", *Electronic Progress* (a Raytheon publication), Vol. 22, No. 3, 1980, pp. 2-9.

[5] Benecke, Th. and A. W. Quick (eds.), *History of German Guided Missiles Development*, AGARD First Guided Missile Seminar, Munich, April 1956.

[6] Müller, Ferdinand, *Leitfaden der Fernlenkung*, Garmisch-Partenkirchen, Deutsche RADAR,1955.

[7] Lusar, Rudolf, *Die deutschen Waffen und Geheimwaffen des 2. Weltkrieges und ihre Weiterentwicklung*, München, J. F. Lehmans, 1956.

[8] Trenkle, Fritz, *Die deutschen Funklenkverfahren bis 1945*, Heidelberg, Alfred Hüthig, 1987.

[9] Benecke, Theodor et al., *Die deutsche Luftfahrt—Flugkörper und Lenkraketen*, Koblenz, Bernard and Graefe, 1987.

[10] Münster, Fritz, "A Guiding System Using Television", in [5], pp. 135-160.

[11] Güllner, Georg, "Summary of the Development of High-Frequency Homing Devices", in [5], pp. 162-172.

[12] Yuan, Luke Chia-Liu, "Homing and Navigational Courses of Automatic Target-Seeking Devices", *RCA Laboratories*, Princeton, N.J., Report No. PTR-12C, Dec. 1943, and *J. Applied Physics*, Vol. 19, 1948, pp. 1122-1128.

[13] Newell, H. E., "Guided Missile Kinematics", *Naval Research Laboratory*, Report No. R-2538, May 1945.

[14] Bennet, R. R. and W. E. Mathews, "Analytical Determination of Miss Distances for Linear Homing Navigation Systems", *Hughes Aircraft Co.*, Culver City, Calif., Report No. TM 260, 1952.

[15] Adler, F. P., "Missile Guidance by Three-Dimensional Proportional Navigation", *J. Appl. Phys.*, Vol. 27, No. 5, 1956, pp. 500-507.

[16] Locke, Arthur S., *Guidance*, a volume in the series *Principles of Guided Missiles Design* edited by Grayson Merrill, Van Nostrand, 1955.

[17] Jerger, J. J., *Systems Preliminary Design*, a volume in the series *Principles of Guided Missiles Design* edited by Grayson Merrill, Van Nostrand, 1960.

[18] Fossier, M. W., "The Development of Radar Homing Missiles", *J. Guidance*, Vol. 7, No. 6, 1984, pp. 641-651.

[19] Anderson, E.W., "Navigational Principles as Applied to Animals", The Duke of Edinburgh Lecture, Royal Institute of Navigation, *J. Navigation*, Vol. 35, No. 1, 1982, pp. 1-27.

[20] Guelman, M., "The Closed-Form Solution of True Proportional Navigation", *IEEE Trans. Aero. Elec. Syst.*, Vol. AES-12, No. 4, 1976, pp. 472-482.

[21] Cochran Jr., J. E., T. S. No, and D. G. Thaxton, "Analytic Solutions to a Guidance Problem", *J. Guidance*, Vol. 14, No. 1, 1991, pp. 117-122.

[22] Dhar, A. and D. Ghose, "Capture Region for a Realistic TPN Guidance Law", *IEEE Trans. Aero. Elec. Syst.*, Vol. 29, No. 3, 1993, pp. 995-1003.

[23] Yuan, Pin-Jar and Jeng-Shing Chern, "Solutions of True Proportional Navigation for Maneuvering and Nonmaneuvering Targets", *J. Guidance*, Vol. 15, No. 1. 1992, pp. 268-271.

[24] Ghose, D., "True Proportional Navigation with Maneuvering Target", *IEEE Trans. Aero. Elec. Syst.*, Vol. 30, No. 1, 1994, pp. 229-237.

[25] Yang, Ciann-Dong, Fang-Bo Yeh, and Jen-Heng Chen, "The Closed-Form Solution of Generalized Proportional Navigation", *J. Guidance*, Vol. 10, No. 2, 1987, pp. 216-218.

[26] Yang, Ciann-Dong, Fei-Bin Hsiao, and Fang-Bo Yeh, "Generalized Guidance Law for Homing Missiles", *IEEE Trans. Aero. Elec. Syst.*, Vol. AES-25, No. 2, 1989, pp. 197-211.

[27] Rao, M. N. "New Analytical Solutions for Proportional Navigation", *J. Guidance*, Vol. 16, No. 3, 1993, pp. 591-594.

[28] Yuan, Pin-Jar and Shih-Che Hsu, "Exact Closed-Form Solution of Generalized Proportional Navigation", *J. Guidance*, Vol. 16, No. 5, 1993, pp. 963-966.

[29] Yuan, Pin-Jar and Shih-Che Hsu, "Rendezvous Guidance with Proportional Navigation", *J. Guidance*, Vol. 17, No. 2, 1994, pp. 409-411.

[30] Baba, Yoriaki, Makoto Yamaguchi, and Robert M. Howe, "Generalized Guidance Law for Collision Courses", *J. Guidance*, Vol. 16, No. 3, 1993, pp. 511-516.

[31] Shukla, U. S. and P. R. Mahapatra, "Generalized Linear Solution of Proportional Navigation", *IEEE Trans. Aero. Elec. Syst.*, Vol. 24, No. 3, 1988, pp. 231-238.

[32] Mahapatra, P. R. and U. S. Shukla, "Accurate Solution of Proportional Navigation for Maneuvering Targets", *IEEE Trans. Aero. Elec. Syst.*, Vol. AES-25, No. 1, 1989, pp. 81-89.

[33] Becker, Klaus, "Closed-Form Solution of Pure Proportional Navigation", *IEEE Trans. Aero. Elec. Syst.*, Vol. 26, No. 3, 1990, pp. 526-533.

[34] Ha, In-Joong, Jong-Sung Hur, Myoung-Sam Ko, and Taek-Lyul Song, "Performance Analysis of PNG Laws for Randomly Maneuvering Targets", *IEEE Trans. Aero. Elec. Syst.*, Vol. 26, No. 5, 1990, pp. 713-721.

[35] Shukla, U. S. and P. R. Mahapatra, "The Proportional Navigation Dilemma—Pure or True?", *IEEE Trans. Aero. Elec. Syst.*, Vol. 26, No. 2, 1990, pp. 382-392.

[36] Ghose, D., "On the Generalization of True Proportional Navigation", *IEEE Trans. Aero. Elec. Syst.*, Vol. 30, No. 2, 1994, pp. 545-555.

[37] Yuan, Pin-Jar and Jeng-Shing Chern, "Ideal Proportional Navigation", *J. Guidance*, Vol. 15, No. 5, 1992, pp. 1161-1165.

[38] Kim, Y. S., H. S. Cho, and Z. Bien, "A New Guidance Law for Homing Missiles", *J. Guidance*, Vol. 8, No. 3, 1985, pp. 402-404.

[39] Yang, Ciann-Dong and Fang-Bo Yeh, "Closed-Form Solution for a Class of Guidance Laws", *J. Guidance*, Vol. 10, No. 4, 1987, pp. 412-415.

[40] Schoen, E. T., "Guidance Laws for Collision Course", RAFAEL, Technical Note 927/TR/3700 (in Hebrew), Jan. 1986.

## FURTHER READING

Murtaugh, S. A. and H. E. Criel, "Fundamentals of Proportional Navigation", *IEEE Spectrum*, Vol. 3, No. 12, Dec. 1966, pp. 75-85.

Stallard, D. V. "Classical and Modern Guidance of Homing Interceptor Missiles", presented to Seminar of Department of Aeronautics and Astronautics, Massachusetts Institute of Technology, *Raytheon Company*, April 1968.

Paarman, L. D., J. N. Faraone, and C. W. Smoots, "Guidance Law Handbook for Classical and Proportional Navigation", *IIT Research Institute, Guidance and Control Information Analysis Center (GACIAC)*, Report No. HB-78-01, 1978.

Zarchan, Paul, *Tactical and Strategic Missile Guidance*, 2nd ed., Washington D.C., AIAA, 1994.

# Chapter 6

# Mechanization of Proportional Navigation

## 6.1 Background

We now leave the problems of kinematics for some time and turn our attention to the question of how PN guidance systems are mechanized and what the main problems involved with the mechnization are. We shall first examine the structure of systems built to mechanize the basic PN laws (5.1)-(5.4). Although the definitions of these laws are compact and simple, the actual systems are not. We shall make the simplifying assumptions made by most authors, as follows.

(i) The engagement is planar.

(ii) Only one control channel is involved in the guidance process, say the $z$-channel, so that angular motions, 'pitch' as it were, are about the $y$-axis of an inertial FOC and displacements are in the plane $y = 0$ of the same FOC. $x$ of course is forward.

## 6.2 On the Structure of PN Systems

It has been stressed several times in this text that guidance systems are by definition feedback systems: at least one feedback loop exists, the guidance loop, where the guidance law plays the main role, usually two or more. The basic PN guidance loop has been shown in Figs. 5.1 and 5.2; more details are now shown in Fig. 6.1. In this figure, the double integrator, the 'comparator' $z_T - z_M$, and the factor $1/r$ belong to the *kinematics* part of the loop. They are, so to speak, supplied by the geometry of the motions and are not parts of any mechanization. It is recalled (see

129

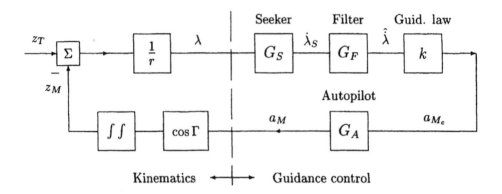

Figure 6.1: Elements of basic PN guidance loop

Sec. 5.4.3) that $\Gamma$ is 0 in TPN and $\delta$ in PPN. The other elements in the figure belong to the *mechanization*, or *guidance control* part of the loop, the input to which is the measured LOS angle $\lambda$. In some cases, the measured range-rate $\dot{r}$ is another input.

Thus, there are three main elements that have to be mechanized, as follows.

(a) The *seeker*, also called *homing head*, is the means for detecting the target T, 'locking' on it, tracking it, and measuring $\dot{\lambda}$. From the point of view of the guidance process, the seeker provides the *measured* $\dot{\lambda}$, denoted $\dot{\lambda}_S$; in some mechanizations, also the (angular) coordinate of T in the seeker frame of coordinates, $e_z$, and the *look angle* $\psi_S$, the angle between the LOS and M's longitudinal axis. For more information on seekers in general, see Sec. 2.5.2; in this chapter, Sec. 6.4.3 deals with some of the problems involved with seekers in PN guidance systems.

Assuming that the behaviour of the seeker is linear, it can be described by a transfer function; the latter is designated $G_S(s)$ and defined as

$$G_S(s) \triangleq \frac{s\lambda_S(s)}{\lambda(s)} ,$$

often approximated by the derivative single-lag transfer function

$$G_S(s) = \frac{s}{1 + sT_S} . \tag{6.1}$$

(b) A *filter* is almost always required, since the $\dot{\lambda}_S$ signal is contaminated by noise. Unless sufficiently fitered, this noise may cause the guidance accuracy to deteriorate . Very often the filter is a single-lag low-pass, i.e., its transfer function is

$$G_F(s) = \frac{1}{1 + sT_F} , \tag{6.2}$$

although in some cases it might possess a more complicated low-pass structure than (6.2). It could be made to depend on time or, for missiles, on dynamic pressure.[1] The output of the filter is often referred to as the *estimated value* of $\lambda$, denoted by $\hat{\lambda}$.

(c) Finally we arrive at the *automatic pilot* (or *autopilot*). This term is of course a misnomer, since the 'autopilot' in Fig. 6.1 and in guidance literature in general actually means the whole lot: for missiles it means aerodynamics, fin servos, inertial and other measuring instruments, amplifiers and filters; and similarly for other guided objects. For an approximation, the autopilot may be represented by the first-order transfer function $G_A(s) = 1/(1 + sT_A)$, by the second-order one

$$G_A(s) \triangleq \frac{a_M(s)}{a_{M_c}(s)} = \frac{1}{1 + 2\zeta\frac{s}{\omega_n} + (\frac{s}{\omega_n})^2} , \qquad (6.3)$$

or by higher-order transfer functions. Actually, of course, the 'constants' $T_A$, $\omega_n$ and $\zeta$ are not really constant, as they depend (for missiles) on speed and altitude.

**Notes.**

(i) If one substitutes $N'v_C$ for $k$ in Fig. 6.1 and equivalent diagrams, the $\cos\Gamma$ element disappears from the loop, since by definition (see (5.24)) $N'v_C = k\cos\Gamma$.

(ii) *Radar seekers* provide, in addition to angle measurements, an estimate of range rate, $\dot{r}$. This makes possible the mechanization of RTPN, described in Sections 5.4.2 and 5.5.1. A great advantage of RTPN is the reduced sensitivity to changes in $v_M$, as will be shown in Sec. 6.4.1(b).[2]

(iii) There exist PN-guided missiles with no seeker on board. This is when *Synthetic PN* is used in combination with command guidance [4]. In such systems, ground radar tracks M and T and measures $\mathbf{r}_M$ and $\mathbf{r}_T$; based on these measurements, the LOS angle rates required for PN are computed, and $\mathbf{a}_{M_c}$ commands are transmitted to M by radio.

## 6.3  The Effects of Dynamics

Throughout Chapter 5 the guided object M was assumed to be ideal, such that there was no delay or lag between $\lambda$ and $a_M$:

$$a_M(t) = a_{M_c}(t) \propto \dot{\lambda}(t) .$$

---

[1]The dynamic pressure is $(1/2)\rho(h)v_M^2$, where $\rho$ is the air density, which depends on the altitude $h$.

[2]While it has not been claimed that bats catch their prey using RTPN guidance, it is a fact that the sonar 'seeker' of certain bats measures range and range-rate as well as azimuth and elevation angles [1-3].

We have already seen three sources of 'dynamics' between $\dot{\lambda}$ and $a_M$, namely the seeker, the filter, and the autopilot (see the last section, (a), (b), and (c), respectively). In this section we shall study the effects that the dynamics of M may have on the guidance process, and in the next section, the effects of some nonlinearities.

Assuming linearity, we define an overall transfer function $G(s)$ such that $G(0) = 1$:

$$G(s) \triangleq \frac{a_M(s)}{ks\lambda(s)} = \frac{\prod_j (1 + sT_j)}{\prod_i (1 + sT_i)} \; , \tag{6.4}$$

where $T_i$, $T_j$ are time constants. By giving $G(s)$ this form we tacitly assume that it has neither integrators nor differentiators. We further assume that the order of the numerator is smaller than the order of the denominator; in fact, in this chapter we shall assume that $G(s)$ has poles only, so that the numerator of (6.4) is 1.

We now proceed by letting $G$ get progressively more and more complex.

## 6.3.1   Single-Lag Dynamics

By definition, $G(s)$ is here

$$G(s) = \frac{1}{1 + sT_M} \; , \tag{6.5}$$

where $T_M$ represents all of the dynamics of M (one may assume, for a first approximation, that $T_M$ equals the sum $T_S + T_F + T_A$). There can, of course, be no simpler dynamics. However, when this dynamics, also expressed by the equation

$$a_M(t) + T_M \frac{da_M(t)}{dt} = a_{M_c}(t) \; ,$$

is combined with the PN guidance law and kinematics, significant analytical difficulties emerge. By the approximative TPN analysis approach of Sec. 5.4.1, where NCC conditions were assumed, the differential equation

$$(1 + T_M \frac{d}{dt})(r\dot{\omega} + 2\dot{r}\omega) = -k\omega$$

is obtained. Doing the same substitutions as in Sec. 5.4.1 one can give this equation the form

$$\tilde{\tau}\tilde{\omega}'' + (3 - \tilde{\tau})\tilde{\omega}' + (N' - 2)\tilde{\omega} = 0$$

where $\tilde{\tau} \triangleq \tau/T_M$, $\tilde{\omega} \triangleq T_M\omega$, and $\tilde{\omega}' \triangleq d\tilde{\omega}/d\tilde{\tau}$.

This equation can be solved in terms of hypergeometric functions [5]. It can be shown that, when $\tilde{\tau} \to 0$, $\tilde{\omega} \propto \tilde{\tau}^{-2}$, provided $N'$ is not a natural number; in other words, the guidance process *diverges*. This conclusion was reached independantly by several authors in the 1950's, using adjoint[3] and numerical methods, where $N'$ could be any (positive) number [6,7,10,12].

---

[3]See the Remark, Sec. 6.5.5.

The detrimental effect of increasung $T_M$, everything else remaining the same, is depicted in Fig. 6.2. In Fig. 6.2(a), miss distances due to heading errors $\gamma_0$ are shown; in Fig. 6.2(b), miss distances due to (constant) target maneuver $a_T$ across the LOS. (It is noteworthy to compare the graphs in these figures with the functions (5.25) and (5.27), respectively, which relate to lag-free dynamics). The graphs of Figure 6.2(b) are obtained from the transfer function

$$\frac{m}{a_T}(s) = \frac{1}{s^2}\left(\frac{s}{s+1/T_M}\right)^{N'}, \tag{6.6}$$

valid for integral values of $N'$; for $N' = 3$ and $N' = 4$, inverse Laplace transforms yield the following miss distance solutions in the time domain.

$$m(t) = \frac{a_T}{2}t_f{}^2\epsilon^{-t_f/T_M} \qquad (N' = 3)$$

$$m(t) = \frac{a_T}{2}t_f{}^2(1 - t_f/3T_M)\epsilon^{-t_f/T_M} \qquad (N' = 4).$$

The effective navigation constant $N'$ and the relative duration of the guidance, i.e., the number $t_f/T_M$, turn out to be important factors. According to a well-known rule of thumb used by designers, this number should be at least 6, many would say 10; see again Fig. 6.2. The same kind of effects is observed when other causes for miss are involved.

Similar conclusions are obtained when *weaving* targets are considered, i.e., when the target maneuver is sinusoidal, given by $a_T(t) = a_{T_p} \sin(\omega t + \phi)$ where $a_{T_p}$ is the amplitude and $\phi$ the phase.[4] Due to the assumed linearity, the resulting miss distance time-history is also sinusoidal (when a certain transient has ended), the amplitude ratio being

$$\left|\frac{m_p}{a_{T_p}}(j\omega)\right| = \left|\frac{1}{(j\omega)^2}\left(\frac{j\omega}{j\omega + 1/T_M}\right)^{N'}\right|$$

$$= T_M^2 \frac{(\omega T_M)^{N'-2}}{[1 + (\omega T_M)^2]^{N'/2}}.$$

The ratio is maximal for $\omega T_M = \sqrt{(N' - 2)/2}$, the value of the maximum being

$$T_M^2 \left(\frac{N' - 2}{2}\right)^{\frac{N'-2}{2}} \left(\frac{N'}{2}\right)^{-\frac{N'}{2}}.$$

For $N' = 3, 4, 5$, numerical values for the maxima are 0.385, 0.250, 0.186, respectively [8]. Thus, for example, if T executes a weave maneuver with an amplitude of

---

[4]In 3-D engagements, weave maneuver is often referred to as barrel roll; see also footnote 1 in Sec. 7.2.

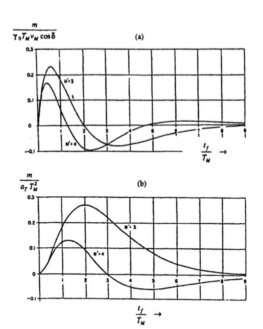

Figure 6.2: miss distances $m$ due to (a) heading error $\gamma_0$, (b) target maneuver across the LOS, $a_T$, as functions of the guidance duration $t_f/T_M$

$50m/s^2$ and M is characterized by $N' = 3$ and $T_M = 1s$, the miss distance may be as high as $0.385 * 50 * 1^2 = 19.2m$ — but may also be close to zero, depending on the phase.

### 6.3.2    Two-Lag Dynamics

We now assume

$$G(s) = \frac{1}{(1 + sT_1)(1 + sT_2)} \, , \tag{6.7}$$

where $T_1$ may represent the dynamics of the seeker and the filter, and $T_2$, the autopilot dynamics.

It has been shown by several authors that the behaviour of the present guidance system does not differ much from that of the single-lag one provided one of the time constants is much smaller than the other. In general, for $T_1$ and $T_2$ such that their sum, say $T_M$, is constant and

$$T_1 = \beta T_M, \quad T_2 = (1 - \beta)T_M, \quad 0 < \beta < 1 \, ,$$

the behaviour of the guidance system deteriorates as $\beta$ increases from 0 to 1/2 (or decreases from 1 to 1/2). These characteristics are well illustrated by the graphs of Fig. 6.3 [6]. Like the graphs of Fig. 6.2, they show miss distances caused by heading errors and target maneuvers, respectively, as functions of the ratio $t_f/T_M$. Note that the rule of thumb regarding this ratio is once more substantiated.

### 6.3.3   Higher-Order Dynamics

Using simulation methods, Garnell has examined PN systems with dynamics having one quadratic lag (i.e., having the structure (6.3)) and two quadratic lags, respectively, and compared the performance obtained with lag-free systems [9]. Generally speaking, the performance of systems with a single quadratic lag was found to be superior to that of systems with two quadratic lags (i.e., fourth order), the lag-free system being the best, needless to say.

When the order of $G(s)$ exceeds 2—some would even say 1—the most practical approach to the analysis and synthesis of PN systems is by using the adjoint method [6,10,12].[5] This has been done by Holt for the third-order dynamics case [13]. In his study, miss distances and lateral accelerations are examined for integral values of $N'$ between 2 and 5. One of the main conclusions is that the best value for $N'$ is 3 if the ratio $v_M/v_T$ exceeds 3.

As a rule, the higher the order of $G(s)$, the lower the performance. For the case of $n$ equal time-constants, $T_M/n$ each, the effect is illustrated in Fig. 6.4, where $m^*$, the highest (i.e., the best from T's point of view) miss distance that T can achieve by certain evasive maneuvers, is shown as a function of the ratio $a_{T_{max}}/a_{M_{max}}$ [14]. Similar results have been obtained when the effects of noise on miss distance were studied [15].

It may be added that the $n = 5$ representation is sometimes considered as especially useful for obtaining preliminary estimates of missile performance [16]; it has even been called "canonical".

### 6.3.4   The Stability Problem

Straightforward analysis of PN guidance loops stability is not in general feasible, due to the variable 'kinematic' gain always present, namely $1/r$. We present two approaches to the study of stability (a third one, of course, is simulation): (a) an intuitive one which provides a condition for stability which, under certain assumptions, is both necessary and sufficient, and (b) an exact one, which provides sufficient conditions.

(a) The *frozen range* method is an intuitive and convenient approach, being based on techniques that many engineers know from their undergraduate studies. By this method, one regards $r$ as constant and analyzes the guidance loop by standard methods, e.g., applying the Nyquist criterion or examining the root locus.

---

[5] See the Remark, Sec. 6.5.5.

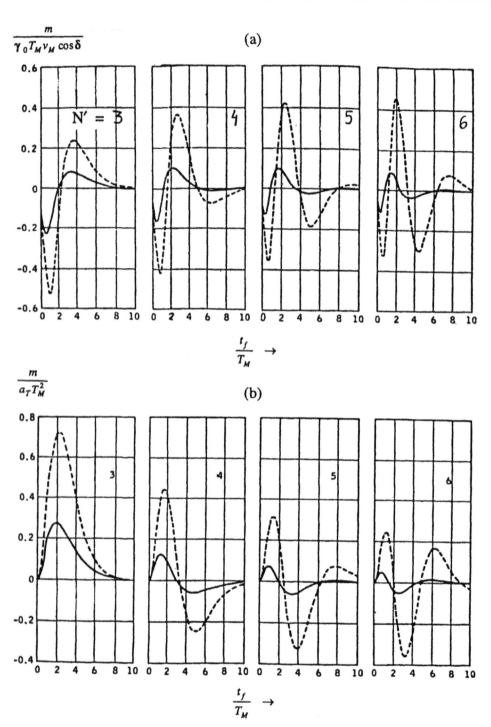

Figure 6.3: Miss distances due to (a) heading error $\gamma_0$, (b) target maneuver across the LOS, $a_T$, as functions of the guidance duration $t_f/T_M$; a single time-constant $T_M$ (solid line) vs. two equal time-constants, $T_M/2$ each (dashed line)

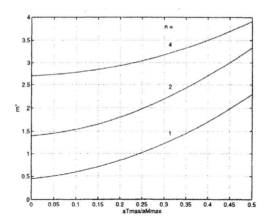

Figure 6.4 Miss distances obtained with dynamics having $n$ equal time-constants, $T_M/n$ each.

Furthermore, analysis of *nonlinear* loops is also possible, using well-known tools such as the frequency-domain describing-function or certain time-domain techniques [18].

Let the open-loop transfer function be $L(s)$; clearly (see Fig. 6.1),

$$L(s) = \frac{1}{r}G_S(s)G_F(s)N'v_C G_A(s)\frac{1}{s^2} , \qquad (6.8)$$

where $r = const.$ and $N'v_C = k\cos\Gamma$. If the order of $1/L(s)$ exeeds 2, instability may arise. The condition for stability is

$$r > r_{cr} \quad (\text{`cr' for `critical'}), \qquad (6.9)$$

where $r_{cr}$ depends on the structure and parameters of $L(s)$. Note that since $G_S(s)$ includes $s$ as a factor, $L(s)$ can be expressed as

$$L(s) = \frac{N'v_C}{r}\frac{G(s)}{s}$$

where $G(s) = [G_S(s)/s]G_F(s)G_A(s)$ and $G(0) = 1$. We recall that the subscripts S, F, and A denote Seeker, Filter, and Autopilot, respectively.

For example, if $G(s)$ is the two-lag transfer function

$$G(s) = \frac{1}{(1 + s\beta T_M)[1 + s(1 - \beta)T_M]} , \quad 0 < \beta < 1 ,$$

then, using standard control theory methods, $r_{cr}$ is easily found to be

$$r_{cr} = N'v_C T_M \beta (1 - \beta) . \tag{6.10}$$

This expression has the maximum value of $N'v_C T_M/4$ for $\beta = 1/2$, thus confirming our statement in Sec. 6.3.2 that $1/2$ is the worst value for $\beta$ from the point of view of guidance.

Instead of using critical range, one can use critical time-to-go. Defining the time-to-go $\tau$ as $r/v_C$, the condition for stabilty is restated as

$$\tau > \tau_{cr} \stackrel{\Delta}{=} \frac{r_{cr}}{v_C} . \tag{6.11}$$

Note that $\tau_{cr}/N'$ depends *only on the dynamics* of M's guidance, i.e., on the time constants present. For the double-lag dynamics example, (6.10) and (6.11) yield

$$\frac{\tau_{cr}}{N'} = T_M \beta (1 - \beta) \le T_M/4 . \tag{6.12}$$

For a second example, let us reexamine the case mentioned in Sec. 6.3.3 for which

$$G(s) = \left( \frac{1}{1 + s\frac{T_M}{n}} \right)^n , \tag{6.13}$$

using the Nyquist stability criterion. It is easily found that the $L(j\omega)$ Nyquist graph crosses the real axis when $\omega = \omega_0$ such that

$$f_0 = \frac{\omega_0}{2\pi} = \frac{1}{4T_M} \frac{\tan(\pi/2n)}{\pi/2n} , \quad L(j\omega_0) = -\frac{N'v_C T_M}{r} \frac{\cos^n(\pi/2n)}{n \tan(\pi/2n)} .$$

Hence the critical range is

$$r_{cr} = -N'v_C L(j\omega_0) = N'v_C T_M \frac{\cos^n(\pi/2n)}{n \tan(\pi/2n)} . \tag{6.14}$$

In the limit, i.e., for $n \to \infty$, $G(s) \to \epsilon^{-sT_M}$, and

$$f_0 = \frac{1}{4T_M} , \quad r_{cr} = \frac{2}{\pi} N'v_C T_M .$$

Fig. 6.5 illustrates the effects of increasing $n$. As $n$ increases, critical range for stability increases, and so does 'sluggishness' (inversely proportional to $f_0$). In the figure, $f_0$ and $r_{cr}$ are made dimensionless by the constants $2\pi/T_M$ and $N'v_C T_M$, respectively.

The effects of increasing $n$ have also been illustrated in Fig. 6.4.

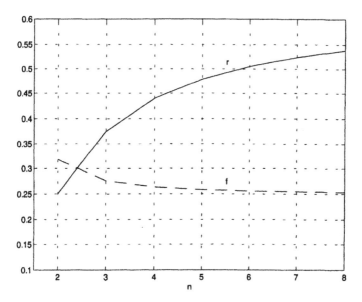

Figure 6.5: Dimensionless cross-over frequency $f_0$ and critical range $r_{cr}$ for systems with $n$ equal time constants, $T_M/n$ each

*Comment.* In spite of the weakness of this approach from the point of view of mathematical rigour ($\dot{r}$ being in fact far from zero), it is a very useful tool for comparative studies.

**(b)** A more sophisticated approach, by Guelman, supplies *sufficient* conditions for *finite-time stability* [19]; it is based on V.M. Popov's stability theory. By this approach, (6.9) is a necessary not sufficient condition. Although the details of the derivation are beyond the scope of this text, the conditions are quite straightforward, being

$$\begin{cases} N' > 2 \\ \tau > -N' Re[\frac{G(j\omega)}{j\omega}] \quad \forall \omega > 0 \ . \end{cases} \tag{6.15}$$

The first of the two conditions should by now be familiar to the reader; the second one bears resemblance to (6.11) but is not equivalent to it.

For example, let $G(s)$ be represented by a product of first-order transfer functions, i.e.,

$$G(s) = \prod_{i=1}^{n} \frac{1}{1 + sT_i} \ , \tag{6.16}$$

where the $T_i$'s could be time constants of $G_S(s)/s$, $G_F(s)$, or $G_A(s)$. It can then be shown that $Re[G(j\omega))/j\omega] > -T_M$ for all $\omega$, where

$$T_M \triangleq \sum_{i=1}^{n} T_i .$$

Hence, a sufficient condition for finite-time stabilty is, by (6.15),

$$\tau > N'T_M .$$

This minimum value for $\tau$ is at least four times as high as the minimum value given in (6.12) as the condition for 'frozen-range' stability for the case $n = 2$. Furthermore, the sufficient condition does not show the effect of the parameter $\beta$ on the behaviour of the system, which the frozen-range condition—and simulation studies—do.

(c) Experience has shown that there is a high correlation between the critical time-to-go (or range) given by (6.11) and (6.15), respectively, and results obtained by numerical simulations and field trials.

## 6.3.5   Conclusions

Sections 6.3.1-6.3.4 indicate that, from the viewpoint of guidance, the dynamics of $G$ should be as fast as possible. In other words, if the dynamics is modelled by the transfer function (6.16) and $T_M$ is as defined above, then $T_M$ should be made as small as possible. However, there are physical limitations—due, e.g., to properties of materials, availability of power and space, environmental conditions— that force in practice a certain minimum value for $T_M$. Furthermore, when radome-refraction errors (see Sec. 6.4.4) and noise inputs (see Sec. 6.5) are also taken into consideration, it often turns out that the optimal value for $T_M$ is not necessarilly the smallest one that engineering makes possible. A sophisticated design might make one or two of the time-constants $T_i$ of (6.16) depend on time or on dynamic pressure; see Sections 6.4.4(e) and 6.5.4(b).

## 6.4   Effects of Nonlinearities in the Guidance Loop

Until now we have been dealing with linear models of M. Real systems, however, have to be described by more complicated models than transfer functions and structures as depicted in Fig. 6.1. In this section we shall assume that M is a missile and examine the following nonlinear phenomena.

(i) Variable missile velocity $v_M$,

(ii) Saturation of lateral acceleration $a_M$,

(iii) Saturations at the seeker,

(iv) Radome refraction error (in radar missiles only),

(v) Imperfect stabilization of the seeker.

By no means is this list complete. We have not included errors in measuring $\lambda$, various errors involved with inertial measurements, effects of sampling, limited capability of fin servo actuators, and some other nonlinear effects. We believe, however, that the items in the list, which are the more important, are sufficient for an introductory text. They will now be treated in Subsections 6.4.1-6.4.5.

## 6.4.1   Variable Missile Speed

Most present-day tactical missiles are propulsed by solid-fuel rocket motors with programmed thrust profiles. The speed, of course, depends not only on the thrust but also on altitude, launch speed (for air-launched missiles), and time history of maneuvers performed.

(a) The first effect of the nonconstant velocity $v_M$ is to prevent M from achieving a straight-line collision course against a nonmaneuvering target. In other words, to cause M to execute unnecessary, 'parasitic' maneuver. In order to clarify the effect, let us refer to parallel navigation, which after all PN is intended to implement. In Sec. 4.2.3 we have obtained a *coupling factor* between $\dot{v}_M$ and $a_M$ for parallel navigation, as follows.

$$\frac{a_M}{\dot{v}_M} = -\tan\delta .$$

When $\delta$ is high, the so-called 'parasitic' lateral acceleration caused by $\dot{v}_M$ may be quite significant. In terms of PN kinematics, the effect can be regarded as if an acceleration bias $\dot{v}_M \sin\delta$ existed. The resulting lateral acceleration $a_M$ would attain—in the case of lag-free dynamics—roughly the same value, or higher if $N' < 4$ (see Sec. 5.4.4(b)).

The effects that the varying speed has on the shape of PN-guided missile trajectories are illustrated by an example summarized in Fig. 6.6. T flies along the straight line $x_T(t) = 2000, y_T(t) = 200\,t$, and $x_M(0) = y_M(0) = 0$. The velocity of M is given by the equation

$$v_M(t) = \begin{cases} 250 + 250\,t, & 0 \leq t < 2 \\ 750\exp[-0.3(t-2)], & t \geq 2 , \end{cases}$$

in $m/sec$ and $sec$, as appropriate. Admittedly this is a difficult velocity profile, since $\dot{v}_M$ varies between $-225$ and $250\,m/sec^2$. At $t = 0$, $\mathbf{v}_M$ is directed at the instantaneous future position of T, i.e., along the direction $\gamma_M(0) = \arcsin(200/250) = 53.1^o$.

Due to the nature of $\dot{v}_M$, first positive and then, from $t = 2$, negative, typical S-formed trajectories result, more pronounced for $N = 3$ than for $N = 4$. For the sake of comparison, the trajectory for parallel navigation ($N = \infty$, as it were) has

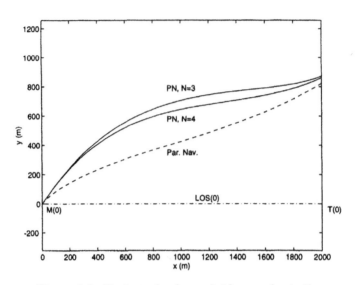

Figure 6.6: Trajectories for variable-speed missiles

been added to the drawing. For each of the three trajectories shown, the highest value of $|a_M|$ exceeds $330\,m/sec^2$; it is highest for the $N = 3$ trajectory, lowest for the parallel-navigation one.

(b) The second effect of the nonconstant velocity $v_M$ is to change the value of $N'$ in missiles where range-rate is not available (e.g., most present-day missiles equipped with electro-optical seekers). In such missiles, the guidance law used is $a_{M_c} = k\lambda$, $k$ being roughly constant. Thus, since $N = k/v_M$, $N' = (k/v_C)\cos\Gamma$, and $\mathbf{v}_C = \mathbf{v}_M - \mathbf{v}_T$, then when $v_M$ changes, so do $N$ and $N'$. It is recalled that the former must be greater than 1 and the latter greater than 2. (Practice has shown that for good results these bounds should be higher; in particular, the latter would be no less than about 3).

When M is not far from a straight-line collision course, such that $|\dot\lambda|$ is small, the equation $v_C = v_M \cos\delta - v_T \cos\theta$ is valid. It follows that $N'$ is then given by the equation

$$N' = \frac{k}{v_M - v_T \frac{\cos\theta}{\cos\delta}}\,,\qquad(6.17)$$

where we have substituted $\delta$ for $\Gamma$, thus having assumed that PPN is the guidance law used. Given $k$, $v_T$, and $v_M$, it is easily seen that $N'$ has its extreme values at the aspect angles $\theta = 0$ and $180^\circ$, for which $\delta \approx 0$ and, therefore, $N' = k/(v_M \mp v_T)$, respectively. When $\theta = \pm 90^\circ$, $N' = k/v_M = N$. (Note that for stationary targets, too, $N' = N$).

For an example, suppose the velocity of a PPN-guided missile can vary between 400 and 800 $m/sec$. In order that $N'$ should remain in the range $3 \leq N' \leq 9$, say, the following inequalities must be satisfied.

$$\frac{k}{800 + v_T} \geq 3, \quad \frac{k}{400 - v_T} \leq 9 .$$

It follows that only tagets with $v_T \leq 100$ $m/sec$ can be tackled at *any* condition, and then, provided $k = 2700$ $m/sec$.

In practice one could try having two distinct values for $k$, a high one, say 3600 $m/sec$, for head-on encounters, and a low one, say 1800 $m/sec$, for tail-on ones. The range of permissible $v_T$ would be zero to 400 $m/sec$ and zero to 200 $m/sec$, respectively.

*Notes.*

(i) Such difficulties are avoidable in radar missiles where $\dot{r}$ is available, so that the RTPN guidance law can be used (see Sec. 5.4.2). When this is done, $v_M$ has no direct effect on the value of $N'$. However, the coupling effect described in (a) still exists.

(ii) The examples given in (a) and (b) above are for ideal, zero-order dynamics, missiles. For missiles with dynamics, say even first-order, the study becomes much more complex [20].

(c) We now turn to three guidance laws specifically developed in order to overcome the problems caused by variable $v_M$.

The first one has in fact already been presented in Sec. 5.5.2(a) as a generalization of TPN.

Another approach is through measuring $\dot{v}_M$ (technically feasible, at least approximately) and compensating for it. The guidance law called *Acceleration compensated PN* (ACPN) does just this [9]. By this law,

$$a_{M_c} = 3v_C \frac{\dot{\lambda}}{\cos \delta} - \dot{v}_M \tan \delta \tag{6.18}$$

where, in the first term, $N'$ has been chosen to be 3, and the second term stems from the coupling factor shown in (a) above.

A third solution has been proposed by Parag and by Gazit and Gutman [21,22]. It is rather similar to ACPN but obtained through different reasoning (see Sec. 7.2.2). They propose the law

$$a_{M_c} = N' \left( v_C \frac{\dot{\lambda}}{\cos \delta} - \frac{\dot{v}_M \tan \delta}{2} \right) , \tag{6.19}$$

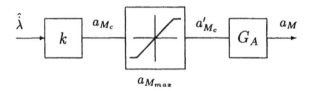

Figure 6.7: Saturation of lateral acceleration

which they call *Terminal guidance* (TG) law. The difference between this law and ACPN is that the acceleration compensation term is $N'/2$ times higher. We shall return to these guidance laws in Sec. 7.2.2.

*Notes.*

(i) Both the second and the third solutions, i.e., the ACPN and TG laws, are PPN laws, such that $\mathbf{a}_{M_c}$ is perpendicular to $\mathbf{v}_M$.

(ii) For both these laws, $\dot{v}_M$ is required; however, in practice it is the componenet of $\dot{\mathbf{v}}_M$ along M's longitudinal body axis that can usually be measured. The error is small except when high angles of attack are involved.

(iii) All three laws belong to the class of *PN Modified by Bias*, more on which will be said in Sec. 7.2.

## 6.4.2   Saturation of Lateral Acceleration

No matter if M is a missile, a boat, or a car, there is always a limit on the lateral acceleration it can attain. For high-altitude missiles, this limit results from angle-of-attack constraints; for low-altitude ones, the limit is due to structural considerations. We have been denoting this limit by $a_{M_{max}}$; when M attains this value it is said to *saturate*. Increasing the $a_{M_{max}}$ capability of any guided object always implies that it is going to cost more, hence the desire to keep the values of $a_M$ required for guidance as low as possible. Indeed, we have seen in previous chapters that this is often the motivation for not utilizing certain guidance laws or for modifying other ones.

A portion of the flow diagram shown in Fig. 6.1 is now redrawn in Fig. 6.7 with the saturation symbol introduced in the loop ahead of the block $G_A$ which represents the autopilot. The symbol means that

$$a'_{M_c} = a_{M_{max}} \, sat(\frac{a_{M_c}}{a_{M_{max}}}) \, , \tag{6.20}$$

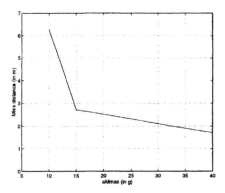

Figure 6.8: The effects of $a_M$ saturation on miss distance

the saturation function $sat(.)$ being defined by

$$sat(u) = \begin{cases} u, & |u| \le 1 \\ sgn(u) = \frac{u}{|u|}, & |u| > 1 \, . \end{cases} \qquad (6.21)$$

Generally speaking, miss distances tend to increase as $a_{M_{max}}$ decreases. This has been illustrated in Fig. 6.4 in the context of high-order dynamics; another illustration, Fig. 6.8, adapted from a study of the performance of a PN-guided missile against an optimally-evading target, also demonstrates this effect [23]. In this diagram, the miss distance $m$ is shown as a function of $a_{M_{max}}$. Note that, for the conditions studied, $m$ is very high when $a_{M_{max}}$ is less than about 10 $g$'s (100 $m/sec$), and that on the other hand there is no point in pushing $a_{M_{max}}$ beyond, say, 20 $g$'s.

For a given maximum desired miss distance, increasing $a_{M_{max}}$ results in increasing launce zones by decreasing minimum launch ranges, and vice versa (these terms have been explained in Sec. 3.3.1). Other limitations, or saturations, that affect the size and shape of launch zones are due to seeker design; they will be dealt with in the next subsection.

### 6.4.3   Saturations at the Seeker

Seekers have already been described in Sec. 2.5.2(d), albeit very briefly. We shall now specifically deal with seekers mounted on missiles (rather than on ground trackers, say).

(a) Due to sensitivity consideration, seekers usually have small fields-of-view (FOV), at most a few degrees; this is one of the reasons why they are usually

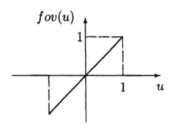

Figure 6.9: Field-of-view saturation

mounted on *gimbals*, such that the forward axis of the seeker can form large angles with the longitudinal axis of M. Various gimballing designs are possible, the most usual one using two mutually perpendicular gimbals (designated yaw and pitch), the outer one of which being orthogonal to the longitudinal axis of M. Sometimes a third gimbal, a roll gimbal, is added as an outer gimbal, coaxial with that axis.

The cone defined by moving the seeker axis about M's longitudinal axis at its largest look-angle is called the *field of regard*. It is recalled that the look angle, denoted by $\psi_S$, is the angle between the seeker axis (i.e., the LOS if one neglects tracking errors) and M's longitudinal axis. Whereas in early ground-to-air and air-to-air missiles the field of regard was about $25^o$ or $30^o$ (half angle), it has increased to $60^o$ and more in modern ones.

When the field of regard does not have to be wide, e.g., in most pure-pursuit mechanizations and in some PN ones as well, one can dispense with the gimbals and mount the seeker directly on the body of M, thus reducing costs significantly. When this is done, the field of regard coincides with the FOV, of course. Such seekers are said to be *strapdown*.[6] We shall return to them later.

Seekers may saturate[7] either because the tracking error $e$ has exceeded the FOV or because the look angle $\psi_S$ has reached the field-of-regard limit $\psi_{S_{max}}$. When either of these saturations has occurred, the guidance loop has effectively become open. There is hardly any hope of successful guidance, i.e., of the miss distance being reasonably small, if the duration of the saturaton was more than a small fraction of the guidance time constant.

The FOV kind of saturation is depicted graphically in Fig. 6.9 and is defined by (6.22), as follows.

$$fov(u) = \begin{cases} u, & |u| \le 1 \\ 0, & |u| > 1. \end{cases} \tag{6.22}$$

---

[6]Since the seeker is 'strapped down', as it were, to M's body.

[7]We are extending the meaning of 'saturation' beyond the conventional one, which is limited to the chatacteristics depicted in Fig. 6.7 and defined by (6.21).

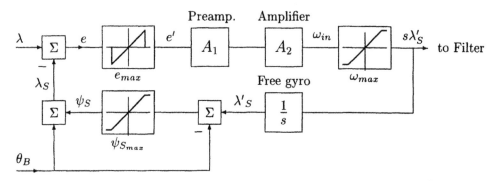

Figure 6.10: Tracking loop, first type

In the block diagram shown in Fig. 6.10,

$$e' = e_{max} fov(\frac{e}{e_{max}}) \, ,$$

$\pm e_{max}$ being the FOV.

(b) Missile seekers (when not strapdown) are always stabilized, which means that they are, at least in principle, completely isolated from the angular motions of M: The seeker axis remains fixed in space even while M pitches, yaws or rolls. This is not a simple task, since the rate which the seeker must measure, $|\dot{\lambda}|$, is usually much smaller than peaks of $|\dot{\theta}_B|$, the body-angle rate. The stabilization is never perfect, and the imperfections may have serious effects on the quality of guidance, as will be shown in Sec. 6.4.5.

There are quite a few ways to implement a stabilized gimballed seeker, all of which being based on a *tracking loop*. We shall examine two common types, without going into technological details but emphasizing general characteristics and saturation problems.

(c) In the first type, the tracking loop has a free gyro in its feedback, mounted on the inner gimbal. (It is recalled that, ignoring a proportionality constant, the transfer function of a free gyro — from the input torque applied to it to the output angle — is ideally $1/s$.) As long as the look angle $\psi_S$ is whithin the field of regard, the 'output' of the gyro is the direction along which the seeker is pointing, i.e., $\lambda'_S$ equals $\lambda_S$ (see Fig. 6.10). Both $\lambda'_S$ and $\lambda_S$, as well as the body attitude angle $\theta_B$, are measured from the same inertial reference line.

The fact that the gyro cannot *precess*, i.e., rotate (about the $y$-axis) faster than at a certain angular speed $\omega_{max}$, is the source of *seeker rate saturation*. A good approximation for this nonlinearity is (see again Fig. 6.10)

$$\dot{\lambda}'_S = \omega_{max} sat(\frac{\omega_{in}}{\omega_{max}}) \, . \tag{6.23}$$

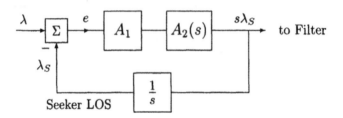

Figure 6.11: Linear tracking loop

In the scheme shown in Fig. 6.10 we have three nonlinearities; as long as none has been reached, the loop behaves just like the linear loop shown in Fig. 6.11, its transfer function being

$$G_S(s) \stackrel{\Delta}{=} \frac{s\lambda_S}{\lambda}(s) = \frac{s}{1 + \frac{s}{A_1 A_2(s)}} \ .$$

This transfer function can often be approximated by the single-lag function

$$G_S(s) = \frac{s}{1 + sT_S} \ , \qquad (6.24)$$

where $T_S = 1/A_1 A_2(0)$.

When rate saturation occurs, i.e., when $|\omega_{in}| > \omega_{max}$, the seeker does not provide correct measurements of $\lambda$, which causes the guidance-loop gain to drop. Also, more significantly, due to this saturation, T is not tracked well, $|e|$ increases, and eventually the FOV-type saturation may cause M to lose T altogether.

(d) In the second type of seeker, the free gyro is replaced by a combination of a rate gyro and a motor. Several mechanizations are possible. In the one shown in Fig. 6.12, from which the $\psi_S$-limitation has been omitted for the sake of clarity, there is an angular-rate loop whithin an angular-displacement one. (Control engineers would refer to such an arrangement as an inner rate loop and an outer position loop.) In such seekers there are encountered in practice only the $e_{max}$ and $\psi_{S_{max}}$ nonlinearities , since $\lambda$-saturation can be avoided by sound design.

*Notes.*
(i) It is recalled that the seeker output, $\dot{\lambda}_S$, must still be filtered in order to obtain $\hat{\dot{\lambda}}$; for example, $s\hat{\lambda} = s\lambda_S/(1 + sT_F)$ for a first-order $G_F$.

(ii) An alternative way of estimating $\dot{\lambda}$, apparently used in most radar seekers, is by making the amplified error $e''$, rather than $\dot{\lambda}_S$, be the seeker output (see Fig. 6.12). From the block diagram, one obtains the transfer function

$$\frac{e''}{\lambda}(s) = \frac{A_1 A_2(s)[1 + sA_3(s)]}{1 + sA_3(s) + A_1 A_2 A_3(s)} \ ;$$

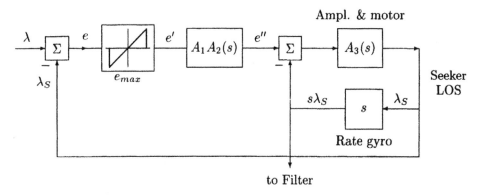

Figure 6.12: Tracking loop, second type

since the loop gains $A_3$ and $A_1 A_2 A_3$ are usually very high, respectively, this transfer function can be approximated by (6.24).

(iii) It has been shown that a better estimate $\hat{\lambda}$ is obtained by either of the sums

$$\hat{\lambda} = \frac{1}{1 + sT_F} \left( s\lambda_S + \frac{s}{1 + sT_F} e \right) ,$$

$$s\hat{\lambda} = \frac{s}{(1 + sT_F)^2} (\lambda_S + e) ,$$

depending on certain parameters of the guidance loop [24]. Both variants stem from the fact that, by definition, $\lambda = \lambda_S + e$, hence $\dot{\lambda} = \dot{\lambda}_S + \dot{e}$.

## 6.4.4   Radome Refraction Error

Most electro-optical missiles have a trasparent sphere at their front end, which serves both as a protection against the surrounding atmosphere and, incidentally, as a part of the optics. From the aerodynamics point of view, this is a bad solution, due to the high drag involved. Radar missiles can dispense with such spherical 'domes' and are protected instead by ogival *radomes* which have much better aerodynamic characteristcs. The *fineness ratio* of such radomes, defined as the length divided by the diameter, $L/D$, is usually between 2 and 3.

The ogival shape of radomes, while reducing drag and therefore increasing ranges, is a source of trouble from the guidance point of view. This results from the fact that, as radiation passes through the radome, it is refracted by a certain angle which, in turn, depends on the look angle. This phenomenon is one of the most dominant contributors to miss distance in radar missiles [25, 26]. In this subsection

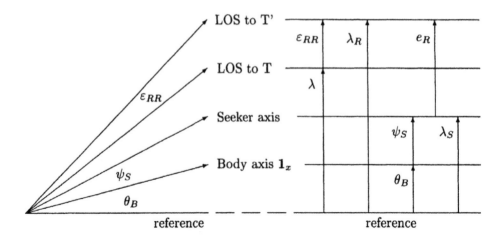

Figure 6.13: Angles involved with radome refraction error

we shall provide some details on the effects of radome refraction on the guidance loop and see what measures can be taken in order to minimize the damage they cause.

(a) For the purpose of analysis, four angles will now be defined, measured from an inertial fixed line (Fig. 6.13).

$\theta_B$, the angle to M's longitudinal body axis,

$\lambda$, the angle to the LOS to the target T,

$\lambda_S$, the angle to the seeker axis,

(these angles have already been defined and used)

$\lambda_R$, the angle to the LOS to the *apparent target* T'.

It follows that the *radome refraction error* $\varepsilon_{RR}$ is given by

$$\varepsilon_{RR} = \lambda_R - \lambda$$

and that the look angle $\psi_S$ equals the difference $\lambda_S - \theta_B$. We shall also use the angle $e_R$, which represents the location of the apparent target in the seeker frame of coordinates and equals the difference $\lambda_R - \lambda_S$.

The *Refraction slope coefficient R* is defined by the equation

$$R \triangleq \frac{d\varepsilon_{RR}}{d\psi_S} \,. \tag{6.25}$$

This definition of course is a local property of the relation $\varepsilon_{RR} = \varepsilon_{RR}(\psi_S)$. Although $R$ is not constant over the whole radome, it may be regarded as constant over a small range of look angles $\psi_S$. Its value is usually no more than a few percent and may be positive or negative.

From the definitions (see also Fig. 6.13), it follows that

$$e_R = \lambda + \varepsilon_{RR} - \lambda_S = \lambda + \varepsilon_{RR} - (\psi_S + \theta_B) .$$

Hence, by (6.25),

$$\dot{e}_R = \dot{\lambda} + R\dot{\psi}_S - (\dot{\psi}_S + \dot{\theta}_B) = \dot{\lambda} - \dot{\theta}_B - (1 - R)\dot{\psi}_S .$$

Since $\psi_S = \lambda_S - \theta_B$, it follows that

$$\dot{e}_R = \dot{\lambda} - (1 - R)\dot{\lambda}_S - R\dot{\theta}_B ,$$

which, since $|R| \ll 1$, may be approximated by the expression

$$\dot{e}_R = \dot{\lambda} - \dot{\lambda}_S - R\dot{\theta}_B .$$

As $\lambda_S + e_R = \lambda_R$ by definition, we finally have

$$\dot{\lambda}_R = \dot{\lambda} - R\dot{\theta}_B . \tag{6.26}$$

(b) We now turn our attention to the effects of radome refraction on the performance of the guidance loop.

Due to the refraction, the seeker input is $\lambda_R$ rather than $\lambda$. Therefore the (unfiltered) output is

$$s\lambda_S(s) = G_S(s)\lambda_R(s) \tag{6.27}$$

where, it is recalled, $G_S(s)$ is the transfer function of the tracking loop such that $G_S(s) \to s$ as $s \to 0$. From aerodynamics we have the transfer function

$$\frac{s\theta_B}{a_M}(s) = \frac{1 + sT_\alpha}{v_M} , \tag{6.28}$$

where $T_\alpha$ is the *turning rate time-constant* of M (see Appendices C.3 and C.4). Combining (6.26), (6.27), and (6.28), the behaviour of the system can be represented by the block diagram shown in Fig. 6.14. This representation should not mislead the reader, however: the effect of radome refraction is through $\dot{\theta}_B$, not $\theta_B$ itself.

The transfer function $a_M(s)/s\lambda(s)$ is easily found to be

$$\frac{a_M(s)}{s\lambda(s)} = \frac{1}{s}\frac{a_M(s)}{\lambda(s)} = \frac{\frac{G_S}{s}G_F kG_A(s)}{1 + \frac{G_S G_F kG_A(s)R(1+sT_\alpha)}{sv_M}} \tag{6.29}$$

(recall that $G_S(s)/s \to 1$ as $s \to 0$). The contribution of radome refraction is clearly the second term in the denominator of (6.29).

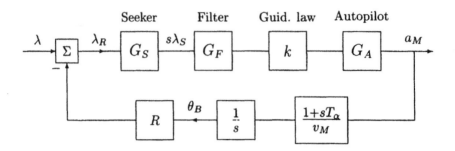

Figure 6.14: Guidance control with radome-refraction present

One can distinguish between static effects, where gains only are involved, and dynamic effects; we shall start with the former.

(c) Due to the refraction, the guidance control gain, ideally $k$, is now

$$k_{RR} = \frac{k}{1 + R\frac{k}{v_M}} \, , \tag{6.30}$$

where the subscript $RR$ signifies 'due to radome refraction'; (6.30) results from (6.29) by letting $s \to 0$. For the sake of brevity we introduce the *refraction coefficient* $\rho$, defined by

$$\rho \triangleq R\frac{k}{v_M} \, , \tag{6.31}$$

so that (6.30) can be rewritten as

$$k_{RR} = \frac{k}{1 + \rho} \, .$$

The change in gain modifies the behaviour of the guidance through the change in the effective navigation constant $N'$. Since $N' = k \cos \Gamma / v_C$, we now have

$$N'_{RR} = \frac{N'}{1 + \rho} = \frac{N'}{1 + \frac{RN' v_C}{v_M \cos \Gamma}} \, . \tag{6.32}$$

Suppose $N' \approx 3$. Then, if $R$ is high (and positive) and $v_C/v_M$ is high too, $N'_{RR}$ may, by (6.32), get dangerously close to 2. If $R$ is negative, $N'$ increases, which results in increasing the detrimental effects of noise (see Sec. 6.5).

(d) The effects on the dynamics would best be shown by assuming for simplicity that $G_S(s)/s, G_F(s)$ and $G_A(s)$ are all single-lag transfer functions with unity gain by definition. Then one obtains from (6.29) the transfer function

$$\frac{a_M(s)}{s\lambda(s)} = \frac{k}{(1 + sT_S)(1 + sT_F)(1 + sT_A) + \rho(1 + sT_\alpha)} \; ,$$

which may be approximated by

$$\frac{k}{1 + \rho + s(T_S + T_F + T_A + \rho T_\alpha)} \tag{6.33}$$

by omitting $s^2$- and $s^3$-terms. We can now rewrite the approximate transfer function of the guidance control in the form

$$\frac{a_M(s)}{s\lambda(s)} = \frac{k}{1 + \rho} \frac{1}{1 + sT_{RR}} \; , \tag{6.34}$$

where $T_{RR}$ is the equivalent time-constant of the guidance control when radome refraction is not negligible. From (6.33) and (6.34) it is evident that

$$T_{RR} = \frac{T_M + \rho T_\alpha}{1 + \rho} \tag{6.35}$$

where

$$T_M \triangleq T_S + T_F + T_A \; .$$

At high altitudes and low speeds, i.e., when dynamic pressure is low, $T_\alpha$ of aerodynamically-controlled missiles may attain high values. If $\rho$ is positive, $T_{RR}$ may then be quite higher than $T_M$, which means that the dynamics of M has become sluggish, with the results known from Sec. 6.3.1; stability problems may result when $\rho$ is negative. See Fig. 6.15(a).

For the fifth-order binomial case, where the nominal guidance-control transfer function is

$$\frac{a_M(s)}{s\lambda(s)} = \frac{1}{(1 + sT_M/5)^5} \; ,$$

it has been shown that radome refraction may cause open-loop instability (i.e., the open-loop transfer-function $a_M(s)/s\lambda(s)$ has at least one pole in the right half-plane) for both positive and negative values of $\rho$ [16].

For an illustration, miss distance as a function of the refraction coefficient $\rho$ is shown in Fig. 6.15(b), adapted from reference [25] (the graph marked 'PN'). The values shown are root-mean-square[8] (see Remark 6.5.5 at the end of this chapter).

---

[8]The root-mean-square (rms) value of $m$, denoted by $m_{rms}$, is the square root of the mean-square value, i.e., $m_{rms} \triangleq \sqrt{E(m^2)}$ where $E(.)$ is the expected value, obtained by statistical methods.

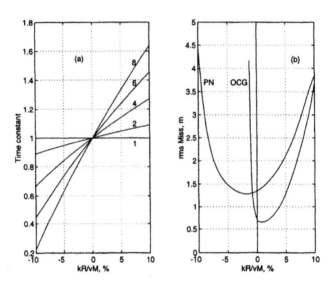

Figure 6.15: (a) The ratio $T_{RR}/T_M$ as a function of radome refraction coefficient $\rho = kR/v_M$, the parameter being $T_\alpha/T_M$; (b) the effect of radome refraction on miss distances

(e) Several lines of action may be considered in order to minimize the effects of radome refraction.

(i) Reduce the fineness ratio $L/D$; this reduces the amount of the refraction, but increases drag. For given radome material and technology of production, $R$ is proportional to $(L/D - 0.5)(\lambda_w/D_{dish})$, where $\lambda_w$ is the radar radiation wavelength and $D_{dish}$ is the antenna dish diameter.[9]

(ii) Reduce the wavelength $\lambda_w$, or rather the ratio $\lambda_w/D_{dish}$.

(iii) In the guidance computer, store a table of values of $R$ vs. look angles, such that the refraction effects can be compensated for during the flight, at least partially. A sophisticated method for doing this has been proposed that uses a compensator based on Kalman filtering [27].

(iv) Increase the system time constant $T_M$; but this of course is undesirable for reasons discussed above in Sections 6.3.1 and 6.3.5.

(v) Compensate by making the filter dynamics (e.g., $T_F$ for a first-order filter) depend on dynamic pressure (see Sec. 6.2(b)).

(vi) Use the effects of imperfect-stabilization of the seeker for partial compensation of negative $R$ (see next subsection).

---

[9]The fineness ratio of a spherical radome is 0.5.

## 6.4.5   Imperfect Stabilization of the Seeker

We have seen that gimballed seekers are gyro-stabilized, in order to separate angular motions of the guided object from those of the LOS. If the stabilization is perfect, the only input to the tracking loop is $\lambda$. However, no stabilization is ever perfect, and in fact parasitic inputs do appear which may have significant effects on the quality of guidance.

(a) As M rotates about its $y$-axis by the angle $\theta_B$, a parasitic input $f(\dot{\theta}_B)$, due to friction in gimbal bearings on one hand and finite stiffness of the stabilization loop on the other hand, enters the tracking loop as a pseudo-$\dot{\lambda}$, as it were. The disturbance is proportional to $f(-\dot{\psi}_S)$, where $\dot{\psi}_S$ is the gimbal angle rate; however, $-\dot{\psi}_S \equiv -\dot{\lambda}_S + \dot{\theta}_B$ can be approximated by $\dot{\theta}_B$, since $|\dot{\lambda}_S| \ll |\dot{\theta}_B|$. For preliminary studies, one may often linearize the function $f(.)$ to have the form $K_{fric}\dot{\theta}_B$, such that the input to the loop is

$$\dot{\lambda}_{fric} = \dot{\lambda} - K_{fric}\dot{\theta}_B \ . \tag{6.36}$$

Comparing (6.36) to (6.26) shows that the effects of imperfect stabilization (as modelled here) are the same as the effects of radome refraction; in order to study them, $K_{fric}$ should be substituted for $R$ and $kK_{fric}/v_M$ for $\rho$ in (6.29)-(6.35).

*Note.* Since the effects of imperfect stabilization of the seeker are similar to those of radome refraction, the former can be used for partial cancellation of the latter when $R < 0$ [28].

(b) Imperfect stabilization exhibits itself in yet another way. In spite of the efforts of designers and manufacturers, there always remains a small *mass-unbalance* in the gimballed seeker and, more significantly, in the gyro. Under an acceleration $a_M$, the mass-unbalance creates a torque which, via the gyro, acts as a disturbance input to the tracking loop. It can easily be shown that the torque is proportional to $a_M$ and to the mass-unbalance.

As an example, let us take the free-gyro seeker described in Sec. 6.4.3(c). Due to the mass-unbalance, a parasitic disturbance $D$ appears as an additional input to the 'block' that represents the gyro, $D$ being directly proportional to the disturbing torque and inversely proportional to the gyro inertia (Fig. 6.16). The tracking loop output is clearly

$$s\lambda_S(s) = \frac{1}{1 + \frac{s}{A_1 A_2(s)}}[s\lambda(s) - D(s)] \ . \tag{6.37}$$

Since $D$ is proportional to $a_M$, say $D = K_g a_M$, we have here effects similar to those caused by radome refraction and by imperfect stabilization as shown in (a), except that there is no $(1 + sT_\alpha)/v_M$ factor in the parasitic feedback from $a_M$. Note that $a_M$ means here 'the component of the total acceleration of M that—through mass-unbalance effects—causes angular motion of the seeker about the $y$-axis'.

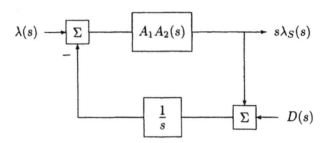

Figure 6.16: Disturbance due to imperfect stabilization

## 6.5   Noise

Sensitivity to noise is known to be Achilles's heel of PN. There are various kinds of noise generally present in PN guidance systems, depending mostly on the type of seeker utilized, i.e., electro-optical or radar and, in the latter case, semiactive or active. As a rule, radar missiles are known to be 'noisier' than electro-optical ones. Nonconstant target maneuvers are also often considered as noise, and will be dealt with later in this section.

Rms miss distances of PN-guided objects are proportional to $\sqrt{\Phi_n}$, where $\Phi_n$ is the noise *equivalent power spectral density (PSD)*. It is related to the intensity, or rms value, $\sigma_n$ of the noise by the equation

$$\Phi_n = 2T_n\sigma_n^2 \ , \tag{6.38}$$

where $T_n$ is the correlation time [15].

We shall distinguish between two kinds of seeker noise, according to the way they interact with the guidance loop, namely *angular noise* and *glint noise* (Fig. 6.17).

### 6.5.1   Angular Noise

Angular noise is usually regarded as the sum of two noises: range-independent noise, sometimes called *fading noise*, and range-dependent one, called *receiver noise*. The former can be caused by many sources, e.g., signal processing effects and gimbal servo inaccuracies. It can also be created externally by a source independent of the radar, such as a distant jammer. The latter is created whithin the receiver itself, e.g., as a result of thermal noise, and target signal must compete with it. It follows from the radar range equation that in semiactive radar its intensity is proportional to the range $r$, and in active radar to $r^2$ [ibid.].

(a) Assuming that the *fading noise* is white, having the equivalent PSD $\Phi_{fn}$ $rad^2/(rad/sec)$, and that the dynamics $G$ of M can be approximated by the first-

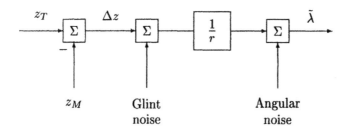

Figure 6.17: Noises in PN guidance loop

order lag $G(s) = 1/(1 + sT_M)$, it has been shown that the rms value of the miss distances that result from fading noise is

$$m_{rms_f} = K_{fn} v_C \sqrt{\Phi_{fn} T_M} \; , \qquad (6.39)$$

where $K_{fn}$ is a constant which is roughly proportional to $N'$ [15]. More generally, for higher-order systems one can write

$$m_{rms_f} = K_{fn} \frac{v_C \sqrt{\Phi_{fn}}}{\sqrt{\omega_b}} \; , \qquad (6.40)$$

where $K_{fn}$ now increases with the order $n$ of the system dynamics $G$, and where the bandwidth $\omega_b$ has been substituted for $1/T_M$ [4, 15].[10] A common approximation for $\omega_b$ is the reciprocal of the sum of the time constants in $G(s)$ (cf. (6.16)). It has been observed that when $G(s)$ has quadratic terms, poor damping may increase miss distances severely [9]. In general, however, miss distances caused by this kind of noise tend to be low.

(b) The rms miss distance that results from white *semiactive radar receiver noise* is given by the equation

$$m_{rms_r} = K_{rn} \frac{v_C^2 \sqrt{\Phi_{rn}}}{\omega_b^{\frac{3}{2}}} \; , \qquad (6.41)$$

---

[10]The bandwidth of $G(j\omega)$ is defined in control engineering context as the angular frequency for which $|G| = 1/\sqrt{2}$ such that $|G(j\omega)| \leq 1/\sqrt{2}$ for all $\omega \geq \omega_b$. For the first-order $G(s) = 1/(1+sT_M)$, $\omega_b T_M = 1$; for the $n$th order $G(s) = 1/(1+sT_M/n)^n$, $\omega_b T_M$ is a monotone function of $n$, increasing from 1 to 2.1 as $n$ increases from 1 to 6; for the quadratic $G(s) = 1/[1 + 2\zeta s/\omega_n + (s/\omega_n)^2]$, $\omega_b/\omega_n = 0.64$, 1, 1.27, 1.43, and 1.54 for $\zeta = 1$, $1/\sqrt{2}$, 1/2, 1/3, and 1/10, respectively. In the context of noise and filtration, another definition for $\omega_b$ usually prevails (see Appendix D.11). However, for the approximate calculations done here, either definition can be used.

where $K_{rn}$ is analogous to $K_{fn}$ and has similar characteristics [15]. A more detailed expression is given by Alpert [4]. For active radar, the exponent of $v_C$ in (6.41) changes to 3, and that of $\omega_b$ to 5/2.

## 6.5.2    Glint Noise

Glint is the most severe noise disturbance to PN radar systems. It results from the motion of the RF-radiation apparent centroid about the target body which, in turn, results from the combined effects of the target reflective elements and the aspect angles; as a matter of fact, the centroid may at times lie outside the physical confines of the target. In the case of aircraft, the measured values of glint usually vary between 1/6 to 1/3 of the maximum span of the aircraft, depending on the distribution of reflecting elements, the aspect angles, and the bandwidth of the angle measurement [29]. The PSD of the centroid motion, $\Phi_g$, is naturally stated in $meter^2/(rad/sec)$, i.e., in units of length rather than angle. An accepted model for this noise (at least for the purpose of preliminary studies) is the spectrum

$$\Phi_g(\omega) = \frac{\Phi_{gn}}{1 + (\frac{\omega}{\omega_g})^2} , \tag{6.42}$$

where $\omega_g$ is the glint half-power frequency and $\Phi_{gn}$ is a constant [ibid.]. Since $\sigma_g^2$, the squared rms value of $\Phi_g$, is the integral $\int_{-\infty}^{\infty} \Phi_g(\omega)d\omega$, it turns out that $\Phi_{gn} = \sigma_g^2/\pi\omega_g$.

Unfortunately, $\omega_g/\omega_b$ is on the order of 1 for most systems, such that the glint noise 'passes' easily through the guidance control: typical values of $\omega_g$ are about $5 - 10 \ rad/sec$. For the purpose of preliminary design, miss distances can be calculated from the equation

$$m_{rms_g} = K_{gn}\sqrt{\Phi_{gn}\omega_b} , \tag{6.43}$$

where $K_{gn}$ is a constant that has the same characteristics as $K_{fn}$ and $K_{rn}$, hence $m_{rms_g}$ is proportional to $\sigma_g\sqrt{\omega_b/\omega_g}$ [4, 15].

*Note.* It has been observed that rms values of miss distance due to glint tend to exceed the rms values of the glint itself, i.e., $m_{rms_g} > \sigma_g$.

## 6.5.3    Target Maneuver

Effects of deterministic target maneuver on the performance of PN- guided objects have been shown in Chapter 5 and earlier in this chapter; see, for example, Eqs. (5.27) and (6.6) and Figs. 5.9, 6.2, 6.3, and 6.4.

Very often, target maneuver is — from the viewpoint of M — of a random nature. One can then regard it as a random disturbance, rather similar to the noise disturbances to the seeker discussed earlier in this section. According to a model which is often used, the target makes step maneuvers with a Poisson distributed starting time. Based on this model, the PSD of the target maneuver 'noise', $\Phi_{tmn}$, equals $4a_T^2/T_P$, where $a_T$ is the level of the maneuver steps and $T_P$ is the average time between maneuvers [4].

The rms miss distance due to this type of noise has been shown to be given by the equation

$$m_{rms_{tm}} = K_{tmn} \frac{\sqrt{\Phi_{tmn}}}{\omega_b^{\frac{5}{2}}}, \qquad (6.44)$$

where $K_{tmn}$ increases with increasing system order and *decreasing* $N'$ [ibid.].

## 6.5.4 Conclusions

(a) In both types of seeker noise models, i.e., angular noise and glint, rms miss distances are roughly proportional to $N'$. This of course explains why designers of PN guidance systems, radar systems in particular, try to keep $N'$ as low as possible (but not too close to 2, of course). However, target maneuver considerations indicate that a 'good' value for $N'$ would be between 4 and 5 [15]. The value most often chosen for radar-seeker systems is about 3, whereas higher values usually prevail in IR-seeker systems.

(b) System design naturally tries to minimize the detrimental effects of noise. This is done by optimizing the radar performance on one hand and judicious design of the system dynamics, particularly the filter $G_F$ (see Sec. 6.2(b) and Fig. 6.1), on the other hand; the latter eventually determines the bandwidth $\omega_b$ of the guidance control system. Whereas angular noise and random target maneuver considerations imply that this bandwidth should be as high as possible, glint noise considerations imply the opposite. The advantages of relatively high bandwidth have been discussed in Sec. 6.3. Based on them, it would seem from (6.40), (6.41), and (6.44) that when $K_{fn}$, $K_{rn}$, and $K_{tmn}$ are high, and high closing velocities were expected, high bandwidth was superior to low one. In practice, of course, a certain compromise, or optimization, must be made: the graph of $m_{rms}$ vs. $\omega_b$ when all types of noise are considered is usually U-shaped.

By an approximative calculation, the optimal value for $\omega_b$ is directly proportional to

$$\left( \frac{K_{tmn}^2 \Phi_{tmn}}{K_{gn}^2 \Sigma\Phi} \right)^{\frac{1}{6}} \qquad (6.45)$$

where $\Sigma\Phi$ is a weighted sum of the seeker noise PSD's; the minimum rms miss distance obtained for this bandwidth is proportional to the product

$$\left( K_{tmn}^2 \Phi_{tmn} K_{gn}^{10} (\Sigma\Phi)^5 \right)^{\frac{1}{12}}. \qquad (6.46)$$

For this optimal condition, $m_{rms_{tm}}$ accounts for 41% of the total rms miss distance [4].

(c) Finally, it should be pointed out that noise inputs increase the rms values of the control acceleration command $a_{M_c}$ required from M.

### 6.5.5   Remark

Most of the results presented in this section have been obtained by the *adjoint* method of approach [6,10,12] which has already been mentioned in this chapter. Since it is not applicable to nonlinear systems, one can resort, when necessary, to another technique, CADET (Covariance Analysis Describing-Function Technique), which combines linear system covariance analysis with multiple-input describing functions [30-33].

A more advanced approach method is through SLAM (Statistical Linearization Adjoint Method), developed by Zarchan, which is a combination of the adjoint method and CADET [34]. Utilizing SLAM, one can rapidly calculate rms miss distances for nonlinear PN guidance systems corrupted by noise.

## 6.6   References

[1] Suga, N., "Biosonar and Neural Computation in Bats", *Scientific American*, Vol. 262, No. 6, June 1990, pp. 34-41 (in some editions, pp. 60-68).

[2] Simmons, J.A., P.A. Saillant, and S.P. Dear, "Through a Bat's Ear", *IEEE Spectrum*, Vol. 29, No. 3, March 1992, pp. 46-48.

[3] Kuc, R. and B. Barshan, "Bat-Like Sonar for Guiding Mobile Robots", *IEEE Cont. Sys. Mag.*, Vol. 12, No. 4, Aug. 1992, pp. 4-12.

[4] Alpert, Joel, "Miss Distance Analysis for Command Guided Missiles", *J. Guidance*, Vol. 11, No. 6, 1988, pp. 481-487.

[5]Guelman, M., "On the Divergance of Proportional-Navigation Homing Systems", *RAFAEL* Report No. 35-137, 1969.

[6]Bennet, R.R. and W.E. Mathews, "Analytical Determination of Miss Distances for Linear Homing Navigation Systems", *Hughes Aircraft Company* Report No. TM 260, 1952.

[7]Jerger, J. J., *Systems Preliminary Design*, a volume in the series *Principles of Guided Missile Design* edited by Grayson Merril, Van Nostrand, 1960.

[8] Ohlmeyer, E. J., "Root-Mean-Square Miss Distance of Proportional Navigation Missile against Sinusoidal Target", *J. Guidance*, Vol. 19, No. 3, 1996, pp. 563-568.

[9] Garnell, P., *Guided Weapon Control Systems*, 2nd ed., Pergamon Press, 1980.

[10] Laning, J. H. and R. H. Battin, "An Application of Analog Computers to the Statistical Analysis of Time-Variable Networks", *Trans. IRE Circuit Theory*, Vol. CT-2, No. 1, 1955, pp. 44-49.

[11] Maney, C. T. (ed.), "Guidance and Control of Tactical Missiles", *AGARD Lecture Series* No. 52, AD-743818, 1972.

[12] Pitman, D. L., "Adjoint Solutions to Intercept Guidance", Paper No. 3C in [11].

[13] Holt, G. C., "Linear proportional navigation: an exact solution for a 3rd order missile system", *Proc. IEE*, Vol. 124, No. 12, 1977, pp. 1230-1236.

[14] Shinar, J. and D. Steinberg, "Analysis of Optimal Evasive Maneuvers Based on Linearized Two-Dimensional Kinematic Model", *Proc. AIAA Guidance and Control Conf.*, 1976, pp. 546-555.

[15] Nesline, F. W. and Paul Zarchan, "Miss Distance Dynamics in Homing Missiles", *Proc. AIAA Guidance and Control Conf.*, 1984, pp. 84-98.

[16] Nesline, F. W. and Paul Zarchan, "Radome Induced Miss Distance in Aerodynamically Controlled Homing Missiles", *Proc. AIAA Guidance and Control Conf.*, 1984, pp. 99-115.

[17] Chadwick, W. R. and P. Zarchan, "Interception of Spiraling Ballistic Missiles", *Proc. Amer. Cont. Conf.*, 1995, pp. 4476-4483.

[18] Bonenn, Z. and N. A. Shneydor, "Divergence Range in a Proportional Navigation Homing System", *Israel Journal of Technology*, vol. 5, No. 1-2, 1967, pp. 129-139.

[19] Guelman, M., "The Stability of Proportional Navigation Systems", *Proc. AIAA Guidance, Navigation and Control Conf.*, 1990, pp. 586-590.

[20] Chadwick, W. R., "Miss Distance of Proportional Navigation Missile with Varying Velocity", *J. Guidance*, Vol. 8, No. 5, 1985, pp. 662-666.

[21] Parag, David, "A PN guidance law for compensating missile acceleration", *RAFAEL* Memorandum, 28 Oct. 1987.

[22] Gazit, R. and S. Gutman, "Development of Guidance Laws for a Variable-Speed Missile", *Dynamics and Control*, Vol. 1, 1991, pp. 177-198.

[23] Julich, P. M. and D. A. Borg, "Effects of Parameter Variations on the Capability of a Proportional Navigation Missile against an Optimally Evading Target in the Horizontal Plane", *Louisiana State University*, Technical Report No. LSU-T-TR-24, AD-700099.

[24] Nesline, F. W. and Paul Zarchan, "Line-of-Sight Reconstruction for Faster Homing Guidance", *J. Guidance*, Vol. 8, No. 1, 1985, pp. 3-8.

[25] Nesline, F. W. and Paul Zarchan, "A New Look at Classical vs Modern Homing Missile Guidance", *J. Guidance*, Vol. 4, No. 1, 1981, pp. 78-85.

[26] Nesline, F. W. and Paul Zarchan, "Missile Guidance Design Tradeoffs for High-Altitude Air-Defense", *J. Guidance*, Vol. 6, No. 3, 1983, pp. 207-212.

[27] Lin, Jium-Ming and Yuan-Fong Chau, "Radome Slope Compensation Using Multiple-Model Kalman Filters", *J. Guidance*, Vol. 18, No. 3, 1995, pp. 637-640.

[28] Fossier, M. W., "The Development of Radar Homing Missiles", *J. Guidance*, Vol. 7, No. 6, 1984, pp. 641-651.

[29] Barton, David K., *Modern Radar System Analysis*, Norwood, Mass., Artech, 1988.

[30] Gelb, A. and R. S. Warren, "Direct Statistical Analysis of Nonlinear Systems - CADET", *AIAA J.*, Vol. 11, No. 5, 1973, pp. 689-694.

[31] Taylor, J. H. and C. F. Price, "Direct Statistical Analysis of Missile Guidance Systems via CADET", *TASC*, Report No. TR-385-2, AD-783098, 1974.

[32] Taylor, J. H., "Handbook for the Direct Statistical Analysis of Missile Guidance Systems via CADET", *TASC*, Report No. TR-385-2, AD-A013397, 1975.

[33] Taylor, J. H. and C. F. Price, "Direct Statistical Analysis of Missile Guidance Systems via CADET", *TASC*, Report No. N76-33259, AD-A022631, 1976.

[34] Zarchan, P., "Complete Statistical Analysis of Nonlinear Missile Guidance Systems — SLAM", *J. Guidance*, Vol. 2, No. 1, 1979, pp. 71-78.

## FURTHER READING

Ivanov, A., "Semi-Active Radar Guidance", *Microwave Journal*, September 1983.

Fossier, M. W., "The Development of Radar Homing Missiles", *J. Guidance*, Vol. 7, No. 6, 1984, pp. 641-651.

Hovanessian, S. A., *Radar System Design and Analysis*, Dedham, Mass., Artech House, 1984.

Wolfe, W.L. and G.J. Zissis (ed.), *The Infrared Handbook*, rev. ed., Washington, D.C., Office of Naval Research, Department of the Navy, 1985.

James, D. A., *Radar Homing Guidance for Tactical Missiles*, Macmillan, 1986.

Maksimov, M. V. and G. I. Gorgonov, *Electronic Homing Systems*, Norwood, MA, Artech House, 1988.

Hovanessian, S. A., *Introduction to Sensor Systems*, Norwood, MA, Artech House, 1988.

Eichblatt, E. J. Jr. (ed.), *Test and Evaluation of the Tactical Missile*, Washington, D. C., AIAA, 1989.

Skolnik, M.I. (ed.), *Radar Handbook*, 2nd ed., McGraw-Hill, 1990.

Lin, C. F., *Modern Navigation, Guidance, and Control Processing*, Prentice-Hall, 1991.

Seyrafi, Khalil and S. A. Hovanessian, *Introduction to Electro-Optical Imaging and Tracking Systems*, Norwood, MA, Artech House, 1993.

Zarchan, Paul, *Tactical and Strategic Missile Guidance*, 2nd ed., Washington D. C., AIAA, 1994.

See also Further Reading for Chapter 5.

# Chapter 7

# Guidance Laws Related to Prop. Navigation

## 7.1 Background

In this chapter we reexamine modifications and extensions of Proportional Navigation (PN) already mentioned in Chapters 5 and 6 and introduce a few new ones.

There is no doubt that PN is an excellent guidance law. The proof is the number of systems that utilize it and the tendency to abandon other laws in its favour. There are, however, some limitations which we will now recall.

PN is the natural guidance law for implementing parallel navigation which, in turn, is optimal for nonmaneuvering targets ($\mathbf{v}_T = const.$), provided $v_M = const.$ When these conditions are not fulfilled, PN can still score a hit, or a near miss, but at a cost: high maneuvers, high control-effort, long time of flight. This is a basic drawback of PN; in Sec. 7.2, several guidance laws motivated by it will be presented.

Next we arrive at problems that arise from mechanizing PN. This guidance law has been shown to be sensitive to noise; two alternative guidance laws, relatively insensitive to a certain kind of noise, are described in Sec. 7.3. A guidance law called *Proportional Lead guidance*, which has some of the characteristics of PN and some of pure pursuit, is presented in Sec. 7.4. Sec. 7.5 deals with guidance laws specifically designed for strapdown seekers. In Sec. 7.6, several examples of *mixed guidance* laws are shown, where PN is combined with some other classical laws, e.g., pure pursuit or LOS guidance.

## 7.2    PN Modified by Bias

Several modifications have been proposed in order to improve the performance of basic PN. Many of the guidance laws that result are in the form of a nonconstant bias added to the well-known PN term $k\dot{\lambda}$ (or $N'v_C\dot{\lambda}$), i.e.,

$$a_{M_c} = k(\dot{\lambda} + b) . \tag{7.1}$$

When the *bias* $b$ is proportional to T's acceleration normal to the LOS, the law is *Augmented PN* (APN), already mentioned in Chapter 5 and to which Sec. 7.2.1 is dedicated; this in fact is the best known Modified PN law [1-6]. The bias $b$ is proportional to M's axial acceleration in the guidance laws described in Section 6.4.1 and in the Guidance-to-Collision laws which will be dealt with in Section 7.2.2. Other laws that belong to the class of *PN Modified by Bias* are as follows.

  * In a law called *Switched Bias PN*, based on sliding-mode control theory, $b$ is proportional to the sign of $\dot{\lambda}$ [7].

  * If $b$ is proportional to $\ddot{\lambda}$, then a rate term, or derivative control, has been added to ordinary PN. We have seen this kind of 'phase-lead compensation' in connection with a LOS guidance law — see Section 2.4.1(b).

  * The Mixed-Guidance law $a_{M_c} = k\dot{\lambda} + k_2\delta$ which will be discussed in Sec. 7.6.1.

  * A similar law, where the lead angle $\delta$ is approximated by the look angle $\psi_S$. (The difference between $\delta$ and $\psi_S$ is the angle of attack $\alpha$; see, e.g., Fig. 3.20).

  * A guidance law for long-range interception of ballistic missiles may have $b$ proportional to the gravitational acceleration halfway between the interceptor and the target [8].

  * In Section 7.3, two more laws that belong to the class of PN Modified by Bias will be presented. In both, $b$ is a nonlinear function of $\dot{\lambda}$.

### 7.2.1    Augmented PN (APN)

First we recall that whereas in parallel navigation $|a_M| \leq |a_T|$ if $K > 1$, in PN the ratio $|a_M/a_T|$ (both accelerations normal to the LOS) increases towards capture, attaining the value $N'/(N'-2)$ in the linear, lag-free ideal case. In real systems the ratio is higher and, more significantly, miss distances are high, too. *Augmented PN* (APN) has been introduced in order to alleviate this difficulty.

In (planar) APN, the acceleration command is the sum

$$a_{M_c} = k\dot{\lambda} + k_a a_T , \tag{7.2}$$

where $a_T$ is the (estimated) acceleration of T normal to the LOS [1-3]. In less general terms, this guidance law is usually defined by the equation

$$a_{M_c} = N'(v_C\dot{\lambda} + \frac{a_T}{2}) . \tag{7.3}$$

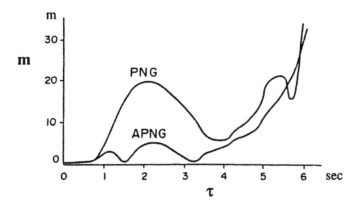

Figure 7.1: Miss distances attained by PN and APN, respectively (Source: Reference 4, Copyright © 1990 AIAA - reprinted with permission)

The term $a_T/2$ in (7.3) becomes clear if the law is stated in terms of the zero-effort miss distance $m$ (see Sec. 5.4.1(b)). When $a_T$ is constant, the zero-effort miss distance $m(t, \tau)$ (in fact, this is the zero $M$-*effort* miss) includes the term $(a_T/2)\tau^2$ in addition to the sum $y(t) + \dot{y}(t)\tau$. Therefore the APN law (7.2) can have the alternative form

$$a_{M_c} = \frac{N'}{\tau^2}m(\tau) = \frac{N'}{\tau^2}[y + \dot{y}\tau + \frac{a_T}{2}\tau^2] \tag{7.4}$$

which, since

$$\dot{\lambda} = \frac{y + \dot{y}\tau}{v_C \tau^2} \ ,$$

is equivalent to (7.3).

For nonplanar engagements, a vector version of the law (7.3) is derived in Appendix D.12.

In APN as presented by Garnell, $k_a$ is a function of $\tau$ [3]. $k$ of course is the PN gain with which we are familiar.

*Note.* (7.2) and (7.3) are TPN-like guidance laws, i.e., $\mathbf{a}_{M_c}$ is normal to the LOS. In the PPN version, i.e., where $\mathbf{a}_{M_c}$ is normal to $\mathbf{v}_M$, the expressions have to be divided by $\cos\delta$.

A simulation study where realistic assumptions were made regarding the dynamics of both M and T shows the miss distance as a function of the time-to-go $\tau_0$ when T initiated its maneuver [4]. In an example shown in Fig. 7.1, drawn from reference [4], APN is seen to achieve a much smaller miss distance than PN for $1.5 < \tau_0 < 3$.

Another simulation study has shown the effectiveness of APN (as compared to

PN) when T carries out a high-g barrel-roll evasive maneuver[1] [5].

In spite of the difficulties involved with estimating $a_T$ (and $\tau$, where required), guidance systems utilizing APN are becoming operational, the US air-defense Patriot probably being one of the first [6]. We shall return to the problem of estimation in Chapter 8. More advanced laws which tackle the problem of maneuvering targets, obtained via the theory of optimal control, will also be dealt with in that chapter.

## 7.2.2   The Guidance-to-Collision Law

This guidance law, proposed by Gazit and Gutman, attains an optimal collision course for a variable-speed missile in much the same way that PN attains such a course for constant speed missiles [9].

(a) The *Guidance-to-Collision* law is defined by the equation

$$\dot{\gamma}_{M_c} = N(\dot{\lambda} + \frac{v_M - \bar{v}_M}{r} \sin \delta) , \quad r \neq 0 , \tag{7.5}$$

where $\bar{v}_M$ is the predicted average speed, i.e., the future distance $S_M$ divided by the time-to-go:

$$\bar{v}_M \triangleq \frac{S_M(\tau)}{\tau} .$$

"This law can be implemented by measuring range, range rate, and missile velocity and by computing the future distance $S_M$ based on some predefined function" [ibid.].

If $\dot{v}_M$ is constant, then (7.5) leads to the guidance law

$$a_{M_c} = N' v_C \left( \frac{\dot{\lambda}}{\cos \delta} - \frac{\dot{v}_M \tan \delta}{2(r\dot{\lambda} \cot \delta - \dot{r})} \right) \tag{7.6}$$

with $\mathbf{a}_{M_c}$ normal to $\mathbf{v}_M$.

(b) The *Terminal guidance* (TG) law of Sec. 6.4.1 is derived from (7.6) by letting $r \to 0$ and substituting $v_C$ for $-\dot{r}$:

$$a_{Mc} = N' \left( \frac{v_C \dot{\lambda}}{\cos \delta} - \frac{\dot{v}_M}{2} \tan \delta \right) . \tag{7.7}$$

Gazit and Gutman examine the performance of an ideal, i.e., linear and lag-free, variable-speed short-range homing missile fired against a crossing, nonmaneuvering target when guided by PN, ACPN (Acceleration Compensated PN, see Sec. 6.4.1(c)) and TG laws, respectively. Typical trajectories and time-histories of lateral acceleration are shown in Fig. 7.2.

---

[1] When T executes this type of maneuver it is effectively advancing on the envelope of a cylinder, its average direction being along the cylinder axis.

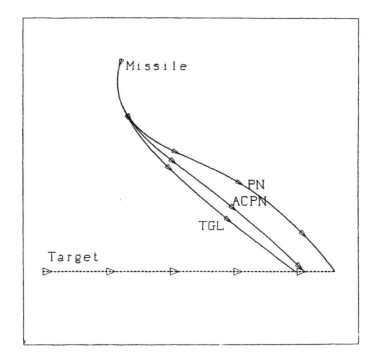

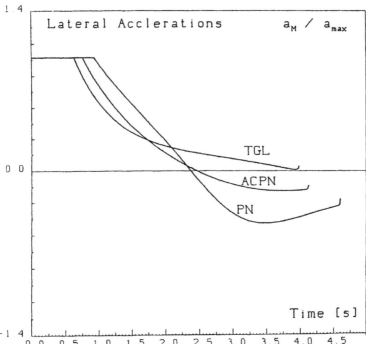

Figure 7.2: Comparison between three guidance laws (Source: Reference 9, with kind permission from Kluwer Academic Publishers)

## 7.3    Guidance Laws for Low LOS Rates

When a PN-guided missile approaches a collision course, the LOS rate $\dot{\lambda}$ drops to low values such that hardly any control $a_M$ is necessary. However, due to noise in the output of the seeker, the *measured* LOS rate $(\dot{\lambda}_S)$ tends to "vacillate between posisitve and negative values, thus [...] imposing severe requirements on the interceptor control system" [10]. In order to reduce this effect two guidance laws have been proposed, wherein nonlinearities are introduced at the input to the guidance control, such that the command is zero for values of $|\dot{\lambda}_S|$ smaller than a certain predetermined threshold. According to the nature of the nonlinearity one distinguishes between *Biased PN* (BPN)[2] and *Dead-Space PN*. Both guidance laws have been considered by Murtaugh and Criel for application in the guidance system of a satellite interceptor [ibid.].

### 7.3.1    Biased PN (BPN)

A bias element is introduced between the seeker and the guidance control, such that its input is $\dot{\lambda}_S$ and its output, $a_{M_c}$. The nonlinear characteristics is (see also Fig. 7.3(a))

$$a_{M_c} = \begin{cases} k(\dot{\lambda}_S - \dot{\lambda}_B) , & \dot{\lambda}_S > \dot{\lambda}_B \\ 0 , & |\dot{\lambda}_S| \leq \dot{\lambda}_B \\ k(\dot{\lambda}_S + \dot{\lambda}_B) , & \dot{\lambda}_S < -\dot{\lambda}_B , \end{cases} \tag{7.8}$$

where $\dot{\lambda}_B > 0$ is the (constant) bias. Using the methods of Sec. 5.4 one can obtain expressions for $\dot{\lambda}(\tau)$, $a_M(\tau)$, and $m(\tau)$. Murtaugh and Criel have shown that the control effort $\Delta v$ required by BPN is $(1 + \dot{\lambda}_B/\dot{\lambda}_0)$ times that reqired by PN, where $\dot{\lambda}_0$ is the initial value of $\dot{\lambda}$. (The control effort, it is recalled, is defined by

$$\Delta v = \int_0^{t_f} |a_M(t)| dt ;$$

it is also called the *velocity increment*, hence the symbol $\Delta v$.)

The problem of the control effort required by BPN when the target maneuvers has been treated by Brainin and McGhee [11,12]. More recently, an analytic study of BPN has been carried out for maneuvering targets and variable closing speeds [13]. The target maneuver is assumed to be proportional to the closing speed. It is shown that the bigger the bias is, the smaller is the capture zone and the higher is the control effort. In another analytic study, BPN is optimized from the point of view of the control effort required; with optimal choice of $\dot{\lambda}_B$, it is possible to effect large savings in control effort required for intercepting maneuvering targets [14].

---

[2]Biased PN should not be confused with the general class of PN Modified by Bias, to which it belongs.

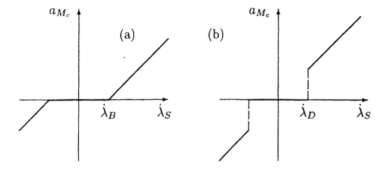

Figure 7.3: Nonlinear characteristics of (a) Biased PN, (b) Dead-Space PN

*Note.* When (7.8) is rewritten as

$$a_{M_c} = k \left[ \dot{\lambda}_S - \dot{\lambda}_B sat \left( \frac{\dot{\lambda}_S}{\dot{\lambda}_B} \right) \right],$$

where $sat(.)$ is the saturation nonlinearity defined in Section 6.4.2, it has the form of the general equation of PN Modified by Bias, (7.1).

## 7.3.2 Dead-Space PN

The nonlinear element that this guidance law utilizes has the characteristics (see also Fig. 7.3(b))

$$a_{M_c} = \begin{cases} k\dot{\lambda}_S, & |\dot{\lambda}_S| > \dot{\lambda}_D \\ 0, & |\dot{\lambda}_S| \le \dot{\lambda}_D \end{cases} \qquad (7.9)$$

where $\dot{\lambda}_D > 0$ is the (constant) dead space. Here, too, the control effort is higher than the control effort required by PN [10]. It seems that this variant of PN has not attracted much research effort the way BPN has.

*Note.* When (7.9) is rewritten as

$$a_{M_c} = k \left[ \dot{\lambda}_S - \dot{\lambda}_D fov \left( \frac{\dot{\lambda}_S}{\dot{\lambda}_D} \right) \right],$$

where $fov(.)$ is the field-of-view nonlinearity defined in Section 6.4.3, it has the form of the general equation of PN Modified by Bias, (7.1).

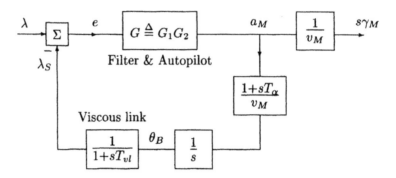

Figure 7.4: Block diagram of Proportional Lead guidance

## 7.4   Proportional Lead Guidance (PLG)

The invention of this guidance law was probably motivated by the sensitivity of PN to noise on one hand and the desire to have relatively economical solutions for air-to-air missile guidance problems, on the other hand [15,16].

In PLG there is actually no tracking loop, or rather, the seeker functions via the missile body itself. The output of the error-measuring device of the seeker is directly delivered to the autopilot (Fig. 7.4). The body causes the gimbal to rotate through a 'viscous link' which has the transfer function $1/(1 + sT_{vl})$. Thus, the transfer function $s\gamma_M(s)/s\lambda(s)$ is

$$\frac{s\gamma_M}{s\lambda}(s) = \frac{1 + sT_{vl}}{1 + s(T_\alpha + \frac{v_M}{G(s)}) + s^2 \frac{v_M T_{vl}}{G(s)}} , \tag{7.10}$$

from which we deduce that the control gain is 1, as in pure pursuit, rather than $N$ as in PN. The term 'proportional lead' is probably due to the lead element $(1 + sT_{vl})$ in the numerator of (7.10). PLG is less sensitive to noise, target accelerations, and wind gusts than PN, but more sensitive to high LOS rates.

## 7.5   Guided Weapons with Strapdown Seekers

The most 'natural' guidance law for guided weapons with strapdown seekers is clearly Pure Pursuit (PP), in fact its Attitude Pursuit variant, with which we dealt in Sec. 3.4. Although this law has been shown to be inferior to Velocity Pursuit which, in turn, is not very satisfactory from kinematics and accuracy points of view, it does have certain advantages, hence its use in several relatively advanced laws to be described in this section.

Strapdown seekers have great technical and economical advantages over gimballed ones [17] (see also Sec. 2.5.2(d)); from the point of view of guidance, however, several difficulties are involved with using them, as follows.

(i) The field of regard is the same as the field of view, hence necessarily limited.

(ii) Inertial LOS rates have to be measured indirectly, by body-mounted gyros.

(iii) Since the characteristics of the error-measuring device (EMD) is not uniform over the whole field of view, the gain of the seeker is variable, depending on the look angle. The error caused by this phenomenon, called the *scale-factor error*, is of high importance in all strapdown systems.

## 7.5.1   An Integral Form of PN

Integrating the basic PN law $\dot{\gamma}_M = N\dot{\lambda}$ one obtains the equation

$$\gamma_M - \gamma_{M_0} = N(\lambda - \lambda_0) , \tag{7.11}$$

where the subscripts '0' denote initial values. This equation represents a law which may be called the *Integral Form of PN*. "An obvious choice of this type of PN would be for a sea skimming missile using a radio altimeter for height control and a good quality free gyro for mid-course heading control. If the end-course guidance is homing, the same gyro can then be used for both pitch and yaw guidance channels" [3].

In particular, a strapdown-seeker version of such a guidance system seems very attractive. Needless to say, there is no PN without a gyro; the present system uses a *body-mounted* free gyro, which can therefore be relatively large and power-consuming — unlike gyros mounted on gimballed-platforms.

A typical mechanization of this guidance law is shown in Fig. 7.5. In this block diagram, $G_2$ and $G_3$ represent filters and amplifiers and $G_4(s) \triangleq \theta_B(s)/\theta_{B_c}(s)$ is the autopilot transfer-function. The autopilot is built in such a way that $G_4(0) = 1$. The same gyro used to provide $\theta_B$ (body angle) feedback to the autopilot is used to provide a signal which combines with the EMD output.

Simple algebra shows that the overall transfer function of the guidance control is

$$\frac{\gamma_M}{\lambda}(s) = \frac{A_1 G_2 G_3 G_4 G_5(s)}{1 + [A_1(s) - G_6(s)A_7(s)]G_2 G_3 G_4(s)} . \tag{7.12}$$

Note that (7.12) has neither poles nor zeros at the origin. Now if the ratio $A_1(s)/A_7(s)$ closely matches $G_6(s)$, then (7.12) becomes

$$\frac{\gamma_M}{\lambda}(s) = A_1 G_2 G_3 G_4 G_5(s) , \tag{7.13}$$

thus implementing the law (7.11), provided the gain of (7.13) is $N$.

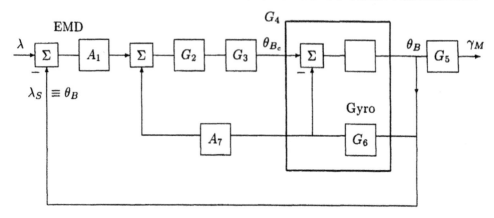

Figure 7.5: Guidance-control of integral form of PN

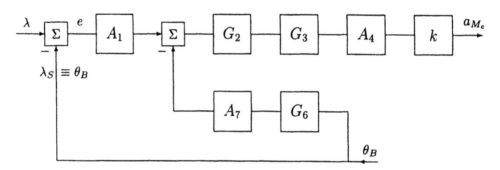

Figure 7.6: Guidance-control of Dynamic Lead Guidance

## 7.5.2    Dynamic Lead Guidance (DLG)

This guidance law has some similarities with both Proportional Lead Guidance of Sec. 7.4 and the integral form of PN described in the preceding subsection. Several versions of DLG have been referred to in the literature; the version presented in Fig. 7.6 provides more details than other versions do [18].

In the block diagram depicted in Fig. 7.6, $G_2$ is the phase-lead filter

$$G_2(s) = \frac{1 + \frac{s}{\omega_1}}{1 + \frac{s}{\omega_2}} , \quad \omega_1 < \omega_2 , \tag{7.14}$$

$G_3$ is the derivative filter

$$G_3(s) = \frac{s}{(1 + s\omega_D)^2} , \tag{7.15}$$

and $A_4$ is an amplifier for "navigation-gain reduction". The output of this amplifier is the input to the PN-gain block with which we are already familiar. Assuming, as in the preceding subsection, that $A_1 \approx A_7 G_6$, it is seen that, at low frequencies, the gain of $a_{M_c}(s)/s\lambda(s)$ is about $A_4 k$, increasing to about $(\omega_2/\omega_1)A_4 k$ at high frequencies. The navigation constant $N$ is therefore $A_4 k/v_M$ and $(\omega_2/\omega_1)A_4 k/v_M$, respectively. $A_4$ and the corner frequencies $\omega_1$ and $\omega_2$ are chosen such that $N \approx 1$ (as required by PP) at low frequencies and $N > 1$ (as required by PN) at higher frequencies. In the said reference, $\omega_1 = 0.2\,rad/sec$, $\omega_2 = 1.0\,rad/sec$.

"Evaluation has indicated that this concept actually has a lower accuracy and lower tolerance to scale factor errors than PN with strapdown seeker" [ibid.]. Nevertheless, this guidance law is said to have been considered for an air-to-ground guided munition and for a homing shell.

# 7.6 Mixed Guidance Laws

Some of the guidance laws that have been described in this chapter have characteristics that are compromises between two of the classical laws. This is achieved by certain changes made to the design of conventional PN. In this section, guidance laws will be presented where the presence of two laws in the same system is more outstanding, hence the term *mixed*.

'Mixing' can be done either by creating a law that combines two basic laws or by having temporal separation between two or more laws. The subsections that follow provide several examples.

## 7.6.1 Mixed Guidance: PP and Parallel Navigation (or PN)

Parallel Navigation, implemented by any of the PN guidance laws, requires high lead-angles $\delta$ when the speed ratio $v_M/v_T$ is low, which is an important disadvantage from the point of view of the seeker; on the other hand, lateral accelerations $a_M$ required are not extremely high. Since PP has the opposite characteristics, a guidance law that has the form

$$a_{M_c} = k_{PN}\dot{\lambda} + k_{PP}\delta \tag{7.16}$$

looks very promising, where $k_{PN}$ and $k_{PP}$ are the gains of the PN and the (velocity-pursuit version of) PP, respectively. In practice one would usually approximate $\delta$ by the gimbal angle $\psi_S$ (see Fig. 3.20), so that the PP law actually becomes attitude pursuit. Also, $k_{PN}$ and $k_{PP}$ need not be constant; they may be made to depend on time, range, or other parameters. For example, $k_{PN}$ would be a function of the speed $v_M$ or the closing speed $v_C$. Furthermore, nonlinear (but probably odd-symmetric) functions of $\psi_S$ may sometimes be preferable to the linear one shown in (7.16).

The law called *Advanced Guidance Law* (AGL), based on modern control theory (see Chapter 8), has been developed for BTT-controlled missiles equipped with strapdown seekers. It is said to be a "Pursuit plus Proportional Navigation type law"; an Extended Kalman Filter is used for estimation of scale-factor errors (see Sec. 7.5 regarding this kind of error) and of the relative position and velocity of T [19].

## 7.6.2    Mixed Guidance: LOS Guidance and Other Laws

In Chapter 2, we have seen that the accuracy of LOS guidance deteriorates as the distance $OM$ between the reference point and M increases. On the other hand, the performance of homing guided objects improves as the range, i.e., the distance $MT$ between M and T, decreases (up to very small ranges). It is therefore natural that guidance systems have been proposed where the the geometrical rule followed switches somewhere along the way from LOS guidance to either PP or parallel navigation (implemented by PN).

An example for the first type of mixing is shown in Fig. 7.7, where a LOS-guided M starts its flight when T is at $\lambda = -30°$. T is flying in the $+y$ direction along the line $x = c$, and the speed ratio is $K = 2$. When the distance $OM$ is $0.8c$, the geometrical rule followed switches over from LOS guidance (full line) to PP (dashed line). Such a mixing was in fact studied as early as in 1955 [20].

The other type of mixing, where guidance is switched from LOS to PN, is in a way easier to implement than the first one. This results from the fact that in LOS guidance there develops 'naturally' a lead-angle $\delta$ which increases with the distance $OM \equiv r_M$ according to the equation

$$\delta = \arcsin\left(\frac{r_M}{K r_T} \sin\theta\right),$$

where $r_T \equiv OT$ (see Sec. 2.3.1(a)). Therefore, if M has been LOS-guided most of the way, until $r_M$ nearly equals $r_T$, switching to PN does not require a large change in $\delta$, which switching to PP does.

This is illustrated in Fig. 7.7, where a parallel navigation straight-line trajectory (dash-dot line) approximates a PN trajectory. The change in $\delta$, hence in $\gamma_M$, required at SO (Switch-Over point) is about $7°$, whereas it is more than $22°$ for the case where switch-over is to PP.

Conceivably, this form of mixed guidance could be suitable for air-to-ship or ship-to-ship guided munitions, as an alternative for another type of mixed guidance, where either *inertial navigation* or a combination of altimeter and gyro heading-control would precede PN. In fact, the 1950's vintage Talos surface-to-air missile utilized LOS guidance for midcourse guidance, and then, for the last 10 seconds or so, "terminal steering placed the missile on a collision course by minimizing the rotation rate of the line of sight"; in other words, guidance switched to PN [21].

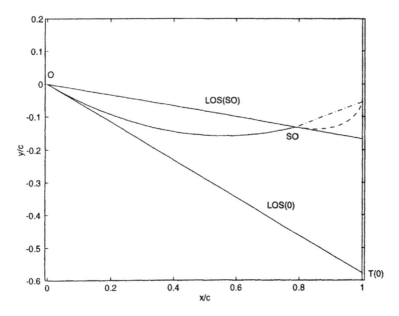

Figure 7.7: Mixed guidance: LOS followed by PP and PN, respectively

## 7.6.3 Combining Midcourse Guidance and PN

Medium-range missiles are, by definition, launched at such long distances from T that throughout most of the flight their seekers are not locked on it. This kind of launch is called *Lock-on after launch* (LOAL), contrasted with *Lock-on before launch* (LOBL), which is the mode of operation of most short-range homing missiles. Clearly, no homing guidance can be accomplished during the first part (usually, the longer one) of the flight of LOAL missiles, lock-on being typically achieved when the range is down to a few kilometers.

In such cases it is advantageous to 'guide' M during that first part in a way that maximizes launch range, for example, or minimizes $t_f$. In the last sentence, the term "guide" has a meaning which is different from the meaning we have given it throughout this text, since there is no geometrical rule that involves the LOS and no feedback loop closed in order to implement it. Still, the process involved is called *midcourse guidance*: When M is sufficiently close to T, guidance switches to *terminal guidance*, or homing, mode, which would be PN, APN, or one of the various 'modern' laws which will be discussed in Chapter 8. (In some texts, the last second or so of the terminal guidance is referred to as the *endgame*, so that the three stages that follow the launch are midcourse guidance, terminal guidance, and endgame). Midcourse guidance is beyond the scope of this book; a few references can be found in references [4,22,23].

## 7.7   References

[1] Siouris, G. M., "Comparison between Proportional and Augmented Proportional Navigation", *Nachrichtentechn. Z.*, Vol. 27, No. 7, 1974, pp. 278-280.

[2] Nesline, F. W. and P. Zarchan, "A New Look at Classical vs Modern Homing Missile Guidance", *J. Guidance*, Vol. 4, No. 1, 1981, pp. 78-85.

[3] Garnell, P., *Guided Weapon Control Systems*, 2nd ed., Pergamon Press, 1980.

[4] Imado, Fumiaki and Takeshi Kuroda, "Optimal Midcourse Guidance for Medium-Range Air-to-Air Missiles", *J. Guidance*, Vol. 13, No. 4, 1990, pp. 603-608.

[5] Imado, Fumiaki and Susumu Miwa, "Missile Guidance Algorithm against High-g Barrel Roll Maneuvers", *J. Guidance*, Vol. 17, No. 1, 1994, pp. 123-128.

[6] Fossier, M. W., "Tactical Missile Guidance at Raytheon", *Electronic Progress, Raytheon Company*, Vol. 22, No. 3, Fall 1980, pp. 2-9.

[7] Ravindra Babu, K., I.G. Sarma, and K.N. Swamy, "Switched Bias Proportional Navigation for Homing Guidance against Maneuvering Targets", *J. Guidance*, Vol. 17, No. 6, 1994, pp. 1357-1363.

[8] Deutsch, O.L., "Interactions between Battle Management and Guidance Law Design for a Strategic Interceptor", in *Agard Lecture Series* LS-173, Sep. 1990.

[9] Gazit, R. and S. Gutman, "Development of Guidance Laws for a Variable-Speed Missile", *Dynamics and Control*, Vol. 1, 1991, pp. 177-198.

[10] Murtaugh, S. A. and H. E. Criel, "Fundamentals of Proportional Navigation", *IEEE Spectrum*, Vol. 3, No. 12, Dec. 1966, pp. 75-85.

[11] Brainin, S. M. and R. B. McGhee, "Analytical Theory of Biased Proportional Navigation", *Hughes Aircraft Company, Aerospace Group*, Report No. TM 665, 1961.

[12] Brainin, S. M. and R. B. McGhee, "Optimal Biased Proportional Navigation", *IEEE Trans. Auto. Contr.*, Vol. AC-13, No. 4, 1968, pp. 440-442.

[13] Yuan, Pin-Jar and Jeng-Shing Chern, "Analytic Study of Biased Proportinal Navigation", *J. Guidance*, Vol. 15, No. 1, 1992, pp. 185-190.

[14] Shukla, U. S. and P. R. Mahapatra, "Optimization of Biased Proportinal Navigation", *IEEE Trans. Aero. Elec. Syst.*, Vol. AES-25, No. 1, 1989, pp. 73-80.

[15] Kuhn, H. L., "Proportional Lead Guidance in a Stochastic Environment" (a Ph.D. thesis), *University of Florida*, 1968.

[16] Kuhn, Harland L., "Proportional Lead Guidance", *U.S. Patent* No. 3,982,714, 28.9.1976.

[17] Vergez, P. L. and J. R. McClendon, "Optimal Control and Estimation for Strapdown Seeker Guidance of Tactical Missiles", *J. Guidance*, Vol. 5, No. 3, 1982, pp. 225-226.

[18] Callen, Thomas R., "Guidance Law Design for Tactical Weapons with Strapdown Seekers", *Proc. AIAA Guidance and Control Conf.*, 1979, pp. 281-293.

[19] Mehra, R. K. and R. D. Ehrich, "Air-to-Air Missile Guidance for Strapdown Seekers", *Proc. IEEE Decision and Control Conf.*, 1984, pp. 1109-1115.

[20] Müller, Ferdinand, *Leitfaden der Fernlenkung*, Garmisch-Partenkirchen, Deutsche RADAR, 1955.

[21] Paddison, F. C., "The Talos Control System", *Johns Hopkins APL Technical Digest*, Vol. 3, No. 2, 1982, pp. 154-156.

[22] Cloutier, J. R., J. H. Evers, and J. J. Feeley, "An Assessment of Air-to-Air Missile Guidance and Control Technology", *Proc. Amer. Contr. Conf.*, 1988, pp. 133-142.

[23] Kumar, R.R., H. Seywald, and E.M. Cliff, "Near-Optimal Three Dimensional Air-to-Air Missile Guidance against Maneuvering Target", *J. Guidance*, Vol. 18, No. 3, 1995, pp. 457-464.

# Chapter 8

# Modern Guidance Laws

## 8.1 Background

The term *modern guidance laws* usually means laws that are based on modern control and estimation theory and technology. This vast scientific discipline encompasses various interconnected technologies, the most prominent of which—for guidance purposes—being optimal-control theory. This theory started to develop in 1959-1961 and aroused great interest among control engineers and theoreticians, not exluding those interested in guidance.

The basic principle of the optimal-control method of design is simplicity itself: one considers a parameter, e.g. the miss distance in guidance problems, and attempts to minimize it. Implementing the principle, however, is sometimes quite complex. The first publications on applying it to problems of guidance seem to date from 1965 [1-3] and 1966 [4]. Since then, hundreds of papers and articles have been published that deal with what has come to be called 'modern guidance' or 'optimal guidance'. Good bibliographies can be found in several reviews [5-8]. Unfortunately, many of the items cited in these reviews have been published as US classified reports and are not easily accessible.

Although most of the references in the literature deal with two-point guidance, the principles of optimal control have been applied to three-point guidance and midcourse guidance as well. In this chapter, however, the term 'modern guidance laws' implies laws applied to two-point guidance.

As is often the case in modern times, theory preceded the technical means necessary for implementing it, notably miniaturized electronics and certain numerical techniques. In fact, by the late 1960's and the early 1970's, the conclusion had been reached that mechanizing advanced guidance laws based on optimal-control theory was not feasible for small tactical weapons. Furthermore, the success of traditional, analog-electronics-based PN-guided weapons in several military conflicts were a negative motivation for applying the advanced laws to new systems.

A change in attitude has come about in the late 1970's, when it was estimated that guided weapons developed at that time might be ineffective against airborne targets of the 1990's (after all, it takes at least a decade or so to develop a new guided weapon system), and the technical difficulties mentioned above had mostly been overcome: sophisticated numerical techniques had been developed for solving the complex equations involved with optimal control, and the modern microprocessor was already in use.

The natural application for modern guidance laws was seen then, and is probably seen today as well, in air defense and air-to-air weapons. This is mainly due to two developments: aircraft have become highly maneuverable, and airborne radar very sophisticated. Development of anti-ballistic missile weapons may also have increased the interest in applying modern control and estimation technologies to missilery.

## 8.2   Methodology

Dealing with 'modern guidance laws', i.e., guidance laws based on optimal-control theory (OCG laws), presents several difficulties which we have not encountered in the preceding chapters of this book.

First of all, one cannot realistically assume that most of the readers of this text are well aquainted with the principles of optimal-control theory.

Another, more basic problem results from the fact that OCG laws cannot be said to implement a simple, straightforward geometrical rule, as is the case with most of the 'classical' guidance laws. (The term *classical guidance laws* usually aplies to laws developed until, say, the early 1960's, i.e., the laws dealt with here in Chapters 1-6). Rather, OCG laws implement a certain mathematical principle.

Furthermore, unlike the classical laws, OCG laws depend on the *detailed dynamics of the guided object* M *and the target* T. In other words, there is no *one* OCG Law.

In view of these difficulties, we will present a review rather than a methodical study. We will advance progressively from the simplest case to more complex ones, roughly following the historical development of the theory, avoiding mathematical details as far as practical.

Thus, in the first part of the chapter, Sec. 8.3, we start with the case where M is linear and lag-free; then we respectively introduce a single lag into M's guidance control, a single lag into the dynamics of T, two lags into M's guidance control, etc. A short summary, Sec. 8.3.6, concludes the first part of the chapter. In the second part, Sections 8.4 and 8.5, more general treatment of OCG is presented. In the third and last part of this chapter, Sec. 8.6 deals with problems involved with the mechanization of OCG, and Sec. 8.7 draws a comparison between OCG and other

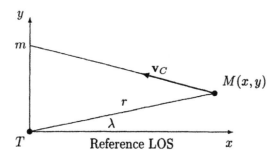

Figure 8.1: Geometry of the simple planar case

guidance laws.

## 8.3 Principles of OCG, and Basic Examples

### 8.3.1 Guidance and Optimal Control

In order to introduce the ideas of optimal control as applied to guidance, let us first examine the very simple planar, linear case already dealt with in Sec. 5.4, where M approaches T at a constant closing velocity $v_C$, near a collision course, such that the following approximations are valid.

$$\begin{cases} r(\tau) = v_C\tau \\ y = r\lambda \ll r \\ a_M = -\ddot{y} \, . \end{cases}$$

(See Fig. 8.1, where T is at the origin, the $x$-axis is along the reference LOS, and M is at $(x, y)$). Earlier in the book, we have called these linearizing approximations the *near collision-course (NCC) assumptions*; we shall use them quite often in the present chapter.

For the first example we also assume that M's dynamics is lag-free, such that $a_M \equiv a_{M_c}$.

We now define a *performance index*, or *cost functional*, $J$ by the equation

$$J = \frac{1}{2}C_1 y^2(t_f) + \frac{1}{2}C_2 \int_{t_i}^{t_f} a_M^2(t)dt \, , \tag{8.1}$$

where $C_1$ and $C_2$ are constants, often called weighting factors, and $t_i$ and $t_f$ are the initial and final moments of the process, respectively. *The essence of optimal control is to try to minimize J.* It is easy to recognize that the two terms in (8.1) represent the miss distance and the control effort, respectively. (The control effort is also sometimes defined as $\int |a_M(t)| dt$ rather than as shown in (8.1)). Thus, a high $C_1/C_2$ ratio stresses the importance of achieving a small miss distance, whereas a high $C_2/C_1$ ratio implies giving more weight to the control effort, i.e., to the energy spent while executing the guidance control. Since only ratios are involved, we can dispense with one factor, say $C_2$, letting it equal 1, so that the dimensions of $J$ and $C_1$ are $m^2 sec^{-3}$ and $sec^{-3}$, respectively.

The *optimal-control problem* is to select a *control* $a_M(t)$ such that $J$ be minimized (a more rigorous statement of the problem will be given in Sec. 8.4). The solution to the present problem has been shown by Bryson [2,9] to be

$$a_M(t) = \frac{3\tau}{\frac{3}{C_1} + \tau^3} [y(t) + \dot{y}(t)\tau] . \tag{8.2}$$

If one requires zero miss at any cost, one lets $C_1 \to \infty$, and the guidance law (8.2) becomes

$$a_M(t) = \frac{3}{\tau^2} [y(t) + \dot{y}(t)\tau] . \tag{8.3}$$

However, from the NCC assumptions it follows that

$$\dot{\lambda} = \frac{d}{dt}\left(\frac{y}{r}\right) = \frac{y(t) + \dot{y}(t)\tau}{v_C \tau^2} \tag{8.4}$$

and, therefore, that the OCG law (8.3) is equivalent to the PN law

$$a_M(t) = 3v_C \dot{\lambda}(t) .$$

In other words, for the guidance problem stated above, *PN with $N' = 3$ is the optimal law* [1,10].

The numerator of (8.4) equals the *zero-effort miss distance* $m(t, \tau)$; we have elaborated on this concept in Sec. 5.4.1(b) in connection with PN. Hence, equivalent forms of the guidance laws (8.2) and (8.3) are, respectively,

$$a_M(t) = \frac{3\tau}{\frac{3}{C_1} + \tau^3} m(t, \tau) \tag{8.5}$$

and

$$a_M(t) = \frac{3}{\tau^2} m(t, \tau) . \tag{8.6}$$

The approach outlined above and, in fact, much of the work published on this topic, is based on a special case of the optimal-control problem, where the dynamics is described by a linear model and the performance index is defined by a quadratic expression. Therefore this problem is generally referred to as the *linear quadratic*, or *LQ*, problem.

Having introduced OCG by this simple example, we now proceed to more complex cases, as follows.

In Sec. 8.3.2, M is still lag-free, but T may be maneuvering.

In Sections 8.3.3 and 8.3.4, M has first- and second-order dynamics, respectively. Higher-order M dynamics is dealt with in Sec. 8.3.5.

A short summary of Sec. 8.3 is presented in Sec. 8.3.6.

## 8.3.2 OCG Laws for a Maneuvering Target

**(a)** Garber studied a *three dimensional* guidance problem and obtained an OCG law for intercepting maneuvering targets by solving the LQ problem with a performance index similar to (8.1), M still being assumed to have zero-order dynamics [11]. The law is stated in terms of the zero-effort miss distance **m**, as follows.

$$\mathbf{a}_M(t) = \frac{3\tau}{\frac{3}{C_1} + \tau^3} \, \mathbf{m}(t, \tau) \, . \tag{8.7}$$

$\mathbf{m}(t, \tau)$ is found from the relative position $\mathbf{r} = \mathbf{r}_T - \mathbf{r}_M$, the target maneuver $\mathbf{a}_T$, the specific thrust $\mathbf{T}$, and the specific drag $\mathbf{D}$ by the equation

$$\mathbf{m}(\tau) = \mathbf{r} + \dot{\mathbf{r}}\tau + \frac{\mathbf{a}_T}{2}\tau^2 + \frac{\dot{\mathbf{a}}_T}{6}\tau^3 - \frac{\mathbf{F}}{2}\tau^2 - \frac{\dot{\mathbf{F}}}{6}\tau^3 \, , \tag{8.8}$$

where $\mathbf{F} \triangleq \mathbf{T} - \mathbf{D}$ and $\dot{\mathbf{a}}_T$ is target acceleration *jerk*. All the vectors in (8.8) are three dimensional, such that we are dealing with a not-necessarily-planar NCC engagement.

If $C_1 \to \infty$ in (8.7), then the guidance law (8.7) reduces to

$$\mathbf{a}_M(t) = \frac{3}{\tau^2} \, \mathbf{m}(t, \tau) \, . \tag{8.9}$$

If, furthermore, $\mathbf{a}_T$ is constant and $\mathbf{F} = \mathbf{0}$, then the guidance law becomes (for planar NCC conditions)

$$a_M = 3\left(\frac{y + \dot{y}\tau}{\tau^2} + \frac{a_T}{2}\right) = 3\left(v_C\dot{\lambda} + \frac{a_T}{2}\right) , \tag{8.10}$$

*which is equivalent to Augmented PN (APN) law with $N' = 3$ (see Sec. 7.2.1 and Appendix D.12)* .

**(b)** By a more realistic model for target maneuver called *exponential* (we are back with the planar, NCC formulation), T maneuvers via a single-lag 'filter' with time constant $T_T$. (For aircraft, the value of $T_T$ is about 3 seconds). The OCG law for this case has been shown by Willems to be

$$a_M(t) = \frac{3\tau}{\frac{3}{C_1} + \tau^3} \left[ y(t) + \dot{y}(t)\tau + a_T(t)T_T^2 \left( \epsilon^{-\frac{\tau}{T_T}} + \frac{\tau}{T_T} - 1 \right) \right] , \tag{8.11}$$

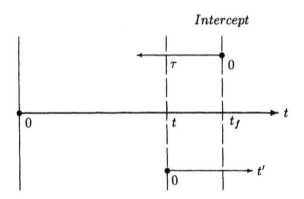

Figure 8.2: Illustrating the definition of $t'$

where $a_T$ is the maneuver of T normal to the LOS [12].

The expression in square brackets in (8.11) equals in fact the zero-effort miss distance $m(t, \tau)$: The first two terms of the sum are already familiar, and the third one represents the contribution of T's maneuver to the predicted miss distance. It can be derived by assuming that the *command* $a_{T_c}$ drops to zero at the moment $t$, such that $a_T$ then starts decaying exponentially:

$$a_T(t') = a_T(t)\epsilon^{-\frac{t'}{T_T}} ,$$

where $t'$ is time that starts anew at $t$ (see Fig. 8.2; note that, by the definitions of $t'$ and $\tau$, $t'$ equals $\tau$ at intercept, when $t = t_f$ ). Two integrations of the function $a_T(t)\epsilon^{-t'/T_T}$ from $t' = 0$ to $t' = \tau$ give the third term in brackets in (8.11).

Therefore we finally have

$$m(t, \tau) = y(t) + \dot{y}(t)\tau + a_T(t)T_T^2 \left( \epsilon^{-\frac{\tau}{T_T}} + \frac{\tau}{T_T} - 1 \right) \tag{8.12}$$

and

$$a_M(t) = \frac{3\tau}{\frac{3}{C_1} + \tau^3} m(t, \tau) .$$

Thus, due to (8.12), the guidance law (8.11) is now seen to be the same as (8.5).

*Notes.*
(i) Both Garber's and Willems's results were published in 1968.
(ii) A more general result has been obtained in 1974 by Asher and Matuszewski, where target-maneuver $a_T(t')$ that starts at $t' = 0$ may be arbitrary [13]: The contribution of T's maneuver to the predicted miss distance is given by the expression

$$m_T(t, \tau) = \int_0^\tau (\tau - t')a_T(t')dt' . \tag{8.13}$$

Using (8.13), a more general expression for $m(t, \tau)$ than (8.12) is

$$m(t, \tau) = y(t) + \dot{y}(t)\tau + m_T(t, \tau) .$$

Of course, since at time $t$ (when, by (8.11), $a_M$ is determined) $a_T(t')$ only starts, one requires knowledge of *future* target acceleration.

This is not a completely unrealistic possibility: for example, if $a_T$ is constant from $t' = 0$ to $t' = \tau$ , then (8.13) gives $m_T(\tau) = (a_T/2)\tau^2$, hence the Augmented PN guidance law which we have seen earlier in this subsection.

For another example, T is an an aircraft that executes a sinusoidal evasive maneuver [14], or T is a penetrating ballistic missile which, due to a certain instability, falls spiralling sinusoidally [15]. If the frequency of the sinusoid, say $\omega$, is known, then it can be shown (see Appendix D.13 for details) that the contribution of T's maneuver to the miss distance can be expressed in terms of its acceleration $a_T(t)$ and jerk $\dot{a}_T(t)$, as follows.

$$m_T(t, \tau) = \frac{1 - \cos \omega\tau}{\omega^2} a_T(t) + \frac{\omega\tau - \sin \omega\tau}{\omega^3} \dot{a}_T(t) . \tag{8.14}$$

(c) We now define the weighting function $f(.)$ which will save us some effort in the rest of this chapter, as follows.

$$f(\xi) \triangleq \frac{e^{-\xi} + \xi - 1}{\xi^2} . \tag{8.15}$$

Using this function we can rewrite (8.11) (for $C_1 \to \infty$) in the form

$$a_M(t) = 3\left[v_C\dot{\lambda} + a_T(t)f(\frac{\tau}{T_T})\right] . \tag{8.16}$$

Since $f(\xi) \to 1/2$ as $\xi \to 0$, we easily deduce that the OCG laws (8.11) and (8.16) approache APN as $\tau/T_T \to 0$.

### 8.3.3   Laws for Systems with 1st Order Dynamics

(a) By assumption, the dynamics of M is given by the transfer function

$$\frac{a_M}{a_{M_c}}(s) = \frac{1}{1 + sT_M} .$$

The OCG law for a maneuvering target ($a_T = const$), still assuming NCC conditions, is now

$$a_{M_c}(t) = \frac{N'(\tau/T_M)}{\tau^2}\left[y(t) + \dot{y}(t)\tau + \frac{a_T}{2}\tau^2 - a_M(t)\tau^2 f(\frac{\tau}{T_M})\right] , \tag{8.17}$$

Table 8.1: The function $N'(\xi)$

| $\xi$ | 7 | 4 | 2 | 1 | 0.5 | 0.2 | 0.1 |
|---|---|---|---|---|---|---|---|
| $N'$ | 4.0 | 5.0 | 7.4 | 12.3 | 22.3 | 52 | 102 |

where $f(.)$ has been defined by (8.15) [16].(Note that in the left-hand side of (8.17) one has $a_{M_c}$, not $a_M$; the two are no more identical.) This law, for which the gain $N'(.)$ is given by the expression

$$N'(\xi) = \frac{6\xi^4 f(\xi)}{3 + 6\xi - 6\xi^2 + 2\xi^3 - 12\xi\epsilon^{-\xi} - 3\epsilon^{-2\xi}} \;,\quad \xi \triangleq \frac{\tau}{T_M} \;, \qquad (8.18)$$

has been obtained assuming $C_1 \to \infty$. $N'(\xi)$ is a monotone-decreasing function of nondimensional time $\xi$, with $N' > 3$ for all $\xi > 0$ and $N' \to 3$ as $\xi \to \infty$. When $\xi$ decreases, $N'$ increases very rapidly, as shown in Table 8.1.

The law (8.17) is sometimes called the *Advanced Guidance Law*. It is said to have been implemented in some missile systems.

*A Note on Notation.* The symbol $N'$ is used for the gain function above (and in similar ones later in this chapter) in order to stress the similarity of the guidance laws to PN. In optimal-control literature, the gain function $\Lambda = N'(\tau)/\tau^2$, called the optimal-control guidance gain, is also often used.

(b) If single lags are present in the controls of *both* M and T, then the OCG law is

$$a_{M_c}(t) = \frac{N'}{\tau^2} \left[ y(t) + \dot{y}(t)\tau + a_T(t)\tau^2 f(\frac{\tau}{T_T}) - a_M(t)\tau^2 f(\frac{\tau}{T_M}) \right] \;, \qquad (8.19)$$

where $N'$ is a function of $\tau$, $C_1$, and $T_M$. For $C_1 \to \infty$, it is the same function as (8.18).

(c) In deriving (8.19) it has implicitly been assumed that T's contribution to the miss distance is via the exponential decay disscussed in Sec. 8.3.2(b). A more general result is

$$a_{M_c}(t) = \frac{N'}{\tau^2} \left[ y(t) + \dot{y}(t)\tau + m_T(t,\tau) - a_M(t)\tau^2 f(\frac{\tau}{T_M}) \right] \;, \qquad (8.20)$$

where $m_T(t,\tau)$ is as defined by (8.13).

## 8.3.4 Laws for Systems with 2nd Order Dynamics

**(a)** Adding another lag to the guidance control, such that

$$\frac{a_M}{a_{M_c}}(s) = \frac{1}{(1 + \frac{s}{\omega_1})(1 + \frac{s}{\omega_2})} , \tag{8.21}$$

makes the solution a little more cumbersome. In order to describe it compactly we shall use state-space formulations.

State variables are defined as follows.

$$\begin{cases} x_1 = y \\ x_2 = \dot{y} \\ x_3 = a_M ; \end{cases}$$

$x_4$ is the command $a_{M_c}$ 'filtered' by the time constant $1/\omega_1$. With no T maneuver, the state equation is

$$\dot{x} = Fx + ga_{M_c} , \tag{8.22}$$

where $x = [x_1 \ x_2 \ x_3 \ x_4]^T$, $g = [0 \ 0 \ 0 \ \omega_1]^T$, and

$$F = \begin{pmatrix} 0 & 1 & 0 & 0 \\ 0 & 0 & -1 & 0 \\ 0 & 0 & -\omega_2 & \omega_2 \\ 0 & 0 & 0 & -\omega_1 \end{pmatrix} . \tag{8.23}$$

Subject to this equation, the solution to the optimal-control problem is

$$a_{Mc} = N'(v_C\dot{\lambda} + G_3 x_3 + G_4 x_4) , \qquad \omega_1 \neq \omega_2 , \tag{8.24}$$

where

$$G_3(\tau) = -f(\omega_2\tau) ,$$

$$G_4(\tau) = \frac{\omega_2}{\omega_1 - \omega_2}[f(\omega_1\tau) - f(\omega_2\tau)] ,$$

$f(.)$ is as defined by (8.15), and $N'$ is a rather complex function of $\tau$, $\omega_1$, $\omega_2$, and $C_1$ [17]. (In control-theory terminology, (8.24) is a full state feedback law).

Several graphs of $N'(\tau)$ are shown in Fig. 8.3 for $\omega_1 = 0.5 \ sec^{-1}$ and several values of $\omega_2$, for (a) $C_1 = \infty$ and (b) $C_1 = 100 \ sec^{-3}$. Note that, in the latter case, $N'(\tau)$ is not monotone.

In the presence of a maneuvering target with a single-lag dynamics as in Sec. 8.3.2(b), the term $G_5 x_5$ is added into the parentheses of (8.24), where $x_5 = a_T$ and $G_5(\tau) = f(\tau/T_T)$.

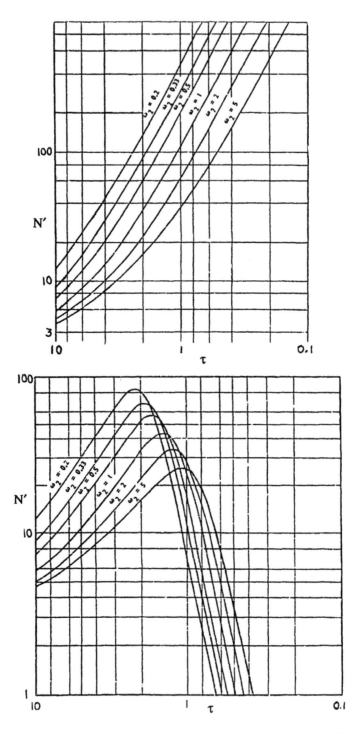

Figure 8.3: The function $N'(\tau)$ for a two-lag system (Source: Reference 17)

For the case $\omega_1 = \omega_2$, see reference [17].

**(b)** The case where $\omega_1$ and $\omega_2$ are complex, i.e., when

$$\frac{a_M}{a_{M_c}}(s) = \frac{1}{1 + 2\zeta\frac{s}{\omega_n} + \left(\frac{s}{\omega_n}\right)^2} , \quad 0 < \zeta < 1, \tag{8.25}$$

has been examined by Stockum and Weimer [18]. In their paper, graphs of $N'(\tau)$ are given for $\zeta = 0.5$ and three values of $\omega_n$, for several values of $C_1$. Some of the functions $N'(\tau)$ are not monotone even when $C_1 = \infty$.

## 8.3.5 Laws for Systems with High-Order Dynamics

OCG laws for systems with dynamics of order higher than 2 become very complex when formulated in the manner shown above. An alternative formulation has been obtained by Rusnak and Meir. They deal with general, high-order autopilot dynamics which may also incorporate a certain constraint on the control $a_{M_c}$, and the guidance laws obtained are in terms of transfer functions and inverse Laplace transforms [19, 20].

Although the Rusnak-Meir OCG law for the *minimum phase* case[1] is relatively compact, its authors conclude that "it does not give improved performance with respect to the first-order approximation".

For *nonminimum* phase autopilots, the exact OCG law found by Rusnak and Meir is complicated to solve, and only "suboptimal/practical" laws are presented by the authors; however, there are situations — combinations of poles and zeros location and acceleration constraint, to be dealt with later — when the full-order guidance law is worth consideration [20, 21].

Examples for third-order dynamics cases where $G(s) =$

$$\frac{1 + 2\zeta_z\frac{s}{\omega_z}k_z + \left(\frac{s}{\omega_z}\right)^2 k_z}{(1 + sT_a)\left[1 + 2\zeta_a\frac{s}{\omega_a} + \left(\frac{s}{\omega_a}\right)^2\right]}, \quad k_z = \pm 1$$

$$\frac{1 + sT_z}{(1 + sT_1)(1 + sT_2)(1 + sT_3)}$$

$$\frac{1 - sT_z}{(1 + sT_1)(1 + sT_2)(1 + sT_3)}$$

are provided in references [19], [20], and [21], respectively.

---

[1]$G(s)$ is minimum-phase if, and only if, all its finite zeros are in the left half of the complex plane; otherwise it is nonminimum phase. The ramp response of $G$ is its response to the input $F(s) = 1/s^2 \longleftrightarrow f(t) = 0$ for $t \leq 0$, $t$ for $t > 0$; if $G$ is minimum-phase, its ramp response is monotone increasing.

Table 8.2: OCG laws discussed in Sec. 8.3

| No. | Subsection | Dynamics of M | $a_T$? | Dynamics of T |
|-----|------------|---------------|--------|---------------|
| 1 | 8.3.1 | lag-free | none | none |
| 2 | 8.3.2(a) | " | yes | lag-free |
| 3 | 8.3.2(b) | " | " | single lag |
| 4 | 8.3.3(a) | singe lag | none | none |
| 5 | 8.3.3(b, c) | " | yes | single lag |
| 6 | 8.3.4 | two lag | none | none |
| 7 | 8.3.4 | " | yes | single lag |
| 8 | 8.3.5 | high order | yes | single lag |

## 8.3.6   A Short Summary

In the preceding five subsections, 8.3.1-8.3.5, a few OCG laws have been presented, derived from the same principle and based on the same assumptions. It seems that a short summary may be helpful at this point.

(a) The laws have been presented in an increasing order of complexity, as listed in Table 8.2.

(b) The assumptions on which the various cases examined have been based are as follows.

(i) The engagement is planar (except in the case of Sec. 8.3.2(a)).

(ii) The geometry is according to the NCC assumptions.

(iii) Where relevant, the accelerations (normal to the LOS) $a_M$ and $a_T$ are known.

(iv) Where relevant, the dynamics of the guidance control is linear and known. Whereas assumption (i) has been made solely for the sake of simplifying the presentation, the other assumptions involve real difficulties which will be dealt with in Sec. 8.6.

(c) Each of the laws can be stated in the form

$$a_{M_c}(t) = \Lambda(\tau)m(t,\tau) \tag{8.26}$$

where, it is recalled, $\Lambda$ is the optimal-control guidance gain and $m$ is the predicted miss distance. $\Lambda(\tau)$ depends, in general, on $C_1$ and on the dynamics of M. $m$ can be expressed in the form

$$m(t,\tau) = y(t) + \dot{y}(t)\tau + m_{M,T}(\tau) , \tag{8.27}$$

where the sum of first two terms is the zero-effort, zero-target-maneuver miss distance and $m_{M,T}$ consists of terms that depend in general on $a_M(t)$ and $a_T(t)$ and on the dynamics of M and T, respectively. For first-order dynamics of M,

$$m_{M,T}(\tau) = m_T(t,\tau) - a_M(t)\tau^2 f(\frac{\tau}{T_M}) , \qquad (8.28)$$

$f(.)$ having been defined in (8.15). $m_T(t,\tau)$ depends on the type of T's maneuver, as shown in Note (ii) in Sec. 8.3.2(b).

For $C_1 \to \infty$, (8.26) can be expressed in the compact form

$$a_{M_c}(t) = \frac{N'(\tau)}{\tau^2}[y(t) + \dot{y}(t)\tau + m_{M,T}(\tau)] , \qquad (8.29)$$

where $N'(\tau)$ depends on the dynamics of M; for lag-free dynamics, $N' = 3$.

(d) (8.29) is a generic guidance law that includes all the OCG laws presented in this Secion for which $C_1 \to \infty$, i.e., the laws that ignore control effort. It belongs to the class of guidance laws called *PN Modified by Bias*, whose form is $a_{M_c} = N'v_C(\dot{\lambda} + b)$ (see Sec. 7.2).

(e) Finally, let us recall that the OCG laws presented in this section and summarized above are based on the performance index (8.1), i.e.,

$$J = \frac{1}{2}C_1 y^2(t_f) + \frac{1}{2}C_2 \int_{t_i}^{t_f} a_M^2(t)dt ,$$

with $C_2$ chosen to be 1. This is not the only performance index possible; in fact, many others have been proposed. For example,

$$J = \frac{1}{2}y^2(t_f) + \frac{1}{2}C_2 \int_{t_i}^{t_f} (t_f - t)^2 a_M^2(t)dt ,$$

$$J = \frac{1}{2} \int_{t_i}^{t_f} m^2 dt ,$$

and

$$J = \frac{1}{2}y^2(t_f) + \frac{1}{2} \int_{t_i}^{t_f} [C_2'm^2 + C_2''(t_f - t)^2 a_M^2(t)]dt ,$$

where $C_2'$ and $C_2''$ are weighting parameters [22].

When minimum fuel consumption (or control effort) is an objective, one could have

$$J = \frac{1}{2} \int_{t_i}^{t_f} |a_M(t)|dt .$$

Design based on this cost functional has been studied and compared with saturating and nonsaturating PN [23].

For applications where it is desired that M intercept T in the shortest time practically possible, performance indices of the form

$$J = C_3 t_f + \frac{1}{2} \int_{t_i}^{t_f} a_M^2(t) dt$$

have been studied [24-26]. In some of these applications, the general, i.e., not necessarily NCC, case is dealt with [25, 26].

Yet other indices will be mentioned in Sec. 8.4.3(b).

## 8.4  A More General Approach to OCG

In Sec. 8.3, the basic principles of OCG have been presented, along with OCG laws for a few cases with varying degrees of complexity. We now proceed to a more general description of the subject [9,27,28].

### 8.4.1  Definitions, and Statement of the Problem

Let $r_M$, $v_M$, and $a_M$ be the position, velocity, and acceleration of M, respectively, relative to an inertial frame of reference. $r_T$, $v_T$, and $a_T$ are similarly defined for T. The relative position, velocity, and acceleration are

$$\begin{cases} \mathbf{r} = \mathbf{r}_T - \mathbf{r}_M \\ \mathbf{v} = \mathbf{v}_T - \mathbf{v}_M \\ \mathbf{a} = \mathbf{a}_T - \mathbf{a}_M \ . \end{cases} \tag{8.30}$$

The system *state* is defined by the $n$-vector $\mathbf{x}$ which has, in general, three components of $\mathbf{r}$, three components of $\mathbf{v}$, and possibly more components, depending on the assumed dynamics of M and T. The *control* of the system is the vector $\mathbf{u}$; in the classical guidance laws discussed in the previous chapters of the book, as well as in the OCG laws of Sec. 8.3, the control was the (negative) command for acceleration, $-\mathbf{a}_{M_c}$.

Terminal states are designated by $t_i$ and $t_f$, respectively, i.e., the engagement starts at $t = t_i$ and ends at $t = t_f$. The subscripts '$i$' and '$f$' signify 'initial' and 'final', respectively. Thus, $x_f = x(t_f)$, $v_f = v(t_f)$, etc.

The dynamical system is represented by the equation

$$\dot{\mathbf{x}} = \mathbf{f}(\mathbf{x}, \ \mathbf{u}, \ t) \tag{8.31}$$

where $\mathbf{f}(.)$ is a vector-valued function of order $n$, in general, nonlinear.

The optimal-control problem is defined in the following manner. Find a control $\mathbf{u}(t)$, $t_i \leq t \leq t_f$, such that the performance index

$$J = \frac{1}{2}g(t_i, t_f, \mathbf{x}_i, \mathbf{x}_f) + \frac{1}{2}\int_{t_i}^{t_f} q(t, \mathbf{x}, \mathbf{u})dt \qquad (8.32)$$

be minimized subject to (8.31) and, possibly, other constraints. $g(.)$ and $q(.)$ are scalar-valued functions. This formulation of the optimal-control problem is very general. Unfortunately there are few conditions under which closed form solutions for $\mathbf{u}(t)$ can be found within its framework.

Disregarding for the moment difficulties of mechanization, there are two important disadvantages to solutions (when they are found) to the optimal-control problem in this formulation , as follows.

Firstly, the solution depends on $t_i$ and/or $t_f$. In other words, for each engagement the solution must be recalculated. Secondly, the solution depends in general on time only, hence it is an *open-loop* solution: it does not depend directly on $\mathbf{x}(t)$. This is quite unlike the classical guidance laws, which are all closed-loop, or *feedback*, laws, where $\mathbf{u}(t)$ depends directly on the system state $\mathbf{x}(t)$. The robustness of feedback systems compared with open-loop ones is well known.[2]

Due to these disadvantages, a less general but more tractable formulation of the optimal-control problem has been developed, in terms of *linear quadratic* (LQ) theory.

## 8.4.2 The LQ Problem

LQ theory is a subset of the general theory which was very briefly outlined in the previous subsection. The difference between the theories can be summarized as follows.

(i) The dynamical system is now assumed to be linear, modeled by the equation

$$\dot{\mathbf{x}}(t) = \mathbf{F}(t)\mathbf{x}(t) + \mathbf{G}(t)\mathbf{u}(t) , \qquad (8.33)$$

where $\mathbf{F}$ and $\mathbf{G}$ are matrices of appropriate dimensions (compare to (8.31) of the previous subsection). It follows from this assumption that NCC conditions prevail.

(ii) The performance index is now quadratic, having the form

$$J = \frac{1}{2}\mathbf{x}_f^T\mathbf{S}\mathbf{x}_f + \frac{1}{2}\int_{t_i}^{t_f}[\mathbf{x}^T(t)\mathbf{Q}(t)\mathbf{x}(t) + \mathbf{u}^T(t)\mathbf{R}(t)\mathbf{u}(t)]dt , \qquad (8.34)$$

where $\mathbf{R}$ is a symmetric positive definite matrix, and $\mathbf{S}$ and $\mathbf{Q}$ are symmetric positive semidefinite matrices (compare to (8.32)).

---

[2]'Robustness' in this context means low sensitivity to modeling errors and to external disturbances.

Ignoring for the moment constraints other than the state equation (8.33), it can be shown [9,27,28] that $J$ is minimized by the control

$$\mathbf{u}(t,\mathbf{x}) = -\mathbf{R}^{-1}(t)\mathbf{G}^{T}(t)\mathbf{P}(t)\mathbf{x}(t) \ , \qquad (8.35)$$

where $\mathbf{P}(t)$ is found by solving the Ricatti nonlinear differential equation

$$-\dot{\mathbf{P}} = \mathbf{P}\mathbf{F} + \mathbf{F}^{T}\mathbf{P} - \mathbf{P}\mathbf{G}\mathbf{R}^{-1}\mathbf{G}^{T}\mathbf{P} + \mathbf{Q} \qquad (8.36)$$

('$(t)$' having been omitted for brevity), subject to

$$\mathbf{P}(t_f) = \mathbf{S} \ . \qquad (8.37)$$

This solution is, indeed, the *guidance law*.

**Remarks.**

(i) If the interceptor for which an OCG law is to be designed is a ground-to-air or air-to-air missile, linearization of the system dynamics is a severe problem. The kinematics in fact is highly nonlinear—we already know this from previous chapters—and so is the aerodynamics.

(ii) The only reason why quadratic forms have been chosen is really mathematical tractability. However, results of designs made according to the LQ approach seem to be satisfactory, at least in simulations; see also the case $\mathbf{Q} = \mathbf{0}$ towards the end of the next subsection.

### 8.4.3   On the Solution to the LQ Problem

(a) The solution $\mathbf{u}(t,\mathbf{x})$, (8.35), has several properties which, when the problem is applied to guidance, make it quite attractive, as follows.

(i) Although $\mathbf{u}(t,\mathbf{x})$ depends on $t_f$, it is independent of both $\mathbf{x}_i$ and $\mathbf{x}_f$, and so is $\mathbf{P}(t)$. This means that the problem need be solved only once (off-line) and that the solution will be valid for all terminal conditions $\mathbf{x}_i$ and $\mathbf{x}_f$.

(ii) $\mathbf{u}(t,\mathbf{x})$ is obviously a direct function of $\mathbf{x}(t)$. This makes $\mathbf{u}$ a feedback control, and the system a feedback system.

(iii) The *control gain* of the system, $\mathbf{C}(t)$, is defined by the equation

$$\mathbf{C}(t) = -\mathbf{R}^{-1}(t)\mathbf{G}^{T}(t)\mathbf{P}(t) \ , \qquad (8.38)$$

such that, by (8.35), the guidance law can now have the compact form

$$\mathbf{u}(t,\mathbf{x}) = \mathbf{C}(t)\mathbf{x}(t) \ . \qquad (8.39)$$

Since $\mathbf{G}$ is assumed to be known, being one of the building blocks of the system model, and $\mathbf{R}$ is a design choice, the information needed for computing $\mathbf{C}(t)$ can be processed off-line and stored in M's guidance computer. However, in practice one must take into account the fact that the system *is* nonlinear; new values of $\mathbf{F}$ and $\mathbf{G}$, and possibly $\mathbf{Q}$ and $\mathbf{R}$, may have to be computed as the engagement proceeds and hence, new values for $\mathbf{C}$.

(b) A problem often ignored in texts that do not specifically deal with design is the choice of the matrices $\mathbf{S}$, $\mathbf{Q}$ and $\mathbf{R}$. Certain guidelines given by Bryson are said to lead to acceptable levels of $x(t_f)$, $x(t)$ and $u(t)$ [9]. By these guidelines, one might choose the matrices to be diagonal, with

$1/S_{ii} =$ max. acceptable value of $[x_i(t_f)]^2$,

$1/Q_{ii} = (t_f - t_i) \times$ max. acceptable value of $[x_i(t)]^2$
   (however, $\mathbf{Q}$ is $\mathbf{0}$ in many designs),

$1/R_{ii} = (t_f - t_i) \times$ max. acceptable value of $[u_i(t)]^2$.

$\mathbf{S}$ is strongly linked to the type of the problem. It very often contains only elements that contribute to the miss distance $m = \|\mathbf{r}_f\|$. In a *rendezvous* problem, however, where the aim is to capture T at zero relative velocity, $\mathbf{S}$ would also include elements that contribute to the value of $\|\mathbf{v}_f\|$; see example (b) in Sec. 8.4.4.

In some applications, e.g., where re-entry vehicles are involved, one may wish to have a constraint on the body attitude-angles at impact, $\theta_{f_y}$ and $\theta_{f_z}$. In planar terms, the performance index $J$ would then have the additional term $(1/2)C_3\theta_f^2$ [29]. Similarly, in applications where high angles of attack should be avoided, a term of the form $(1/2)C_4\alpha_f^2$ would be added to the performance index. Other parameters that might be added could result from seeker FOV limitations, autopilot considerations, and state estimate uncertainty [30, 31].

When $\mathbf{Q} = 0$, the second term of (8.34), which now includes only the integral of $\mathbf{u}^T(t)\mathbf{R}(t)\mathbf{u}(t)$, is quite properly called the *control effort*. The choice of the elements of $\mathbf{R}$ depends on the application and on the structure of the guidance control. For example, it has been pointed out that, since specific drag $D$ depends on the lateral maneuver of M, $a_M$, roughly according to the quadratic expression

$$D = D_0 + D_1 a_M^2 ,$$

and velocity loss is approximately given by

$$\Delta v = \int_{t_i}^{t_f} D(t)dt ,$$

a natural choice for $\mathbf{u}$ would be the two lateral components of $\mathbf{a}_M$, and for $\mathbf{R}$ the unity matrix $\mathbf{I}$ weighted by an appropriate scalar [3]. This physical rationale also adds a justification to the quadratic nature of the performance index.

## 8.4.4   Two Examples

The first example will show how the guidance laws obtained in Sections 8.3.1 and 8.3.2 for lag-free dynamics are derived from the general case presented in Sec. 8.4.2.

The second one will deal with a rendezvous problem.

(a) A planar engagement is assumed; this assumption, made solely for simplicity, makes it possible to define a state vector $\mathbf{x}$ with four rather than six components, as follows.

$x_1$ and $x_2$ are the components of $\mathbf{x}$, and $x_3$ and $x_4$ the components of $\mathbf{v}$, in the $x$ and $y$ axes of an inertial frame of coordinates (IFOC), respectively. Consequently, $\dot{x}_3$ and $\dot{x}_4$ are the components of $\mathbf{a}$ in the same IFOC, respectively.

We now further assume that T does not maneuver, i.e., $\mathbf{a}_T = \mathbf{0}$, hence $\mathbf{a} = -\mathbf{a}_M$, and recall that the dynamics of M is lag-free such that $\mathbf{a}_M = \mathbf{a}_{M_c}$. By defining the control $\mathbf{u}$ to be $-\mathbf{a}_{M_c}$, the linear model of the system is

$$\dot{\mathbf{x}} = \mathbf{F}\mathbf{x} + \mathbf{G}\mathbf{u} = \begin{pmatrix} \mathbf{0} & \mathbf{I} \\ \mathbf{0} & \mathbf{0} \end{pmatrix} \mathbf{x} + \begin{pmatrix} \mathbf{0} \\ \mathbf{I} \end{pmatrix} \mathbf{u} , \tag{8.40}$$

where $\mathbf{I}$ is a $2 \times 2$ identity matrix, $\mathbf{0}$ is a $2 \times 2$ zero matrix, and $\mathbf{u} = [u_1 \ u_2]^T$. The matrices $\mathbf{S}$, $\mathbf{Q}$ and $\mathbf{R}$ of the performance index (8.34) are

$$\mathbf{S} = C_1 \begin{pmatrix} \mathbf{I} & \mathbf{0} \\ \mathbf{0} & \mathbf{0} \end{pmatrix}, \quad \mathbf{Q} = 0, \quad \mathbf{R} = \begin{pmatrix} 1 & 0 \\ 0 & 1 \end{pmatrix} , \tag{8.41}$$

hence, by substitution into (8.34),

$$J = \frac{1}{2}C_1[x_1^2(t_f) + x_2^2(t_f)] + \frac{1}{2}\int_{t_i}^{t_f}[u_1^2(t) + u_2^2(t)]dt . \tag{8.42}$$

By (8.35) and (8.36), the solution to the present problem is

$$\mathbf{u}(t, \mathbf{x}) = -\mathbf{R}^{-1}(t)\mathbf{G}^T(t)\mathbf{P}(t)\mathbf{x}(t) = -\begin{pmatrix} \mathbf{0} & \mathbf{I} \end{pmatrix} \mathbf{P}(t) \mathbf{x}(t) \tag{8.43}$$

with

$$-\dot{\mathbf{P}} = \mathbf{P}\begin{pmatrix} \mathbf{0} & \mathbf{I} \\ \mathbf{0} & \mathbf{0} \end{pmatrix} + \begin{pmatrix} \mathbf{0} & \mathbf{0} \\ \mathbf{I} & \mathbf{0} \end{pmatrix}\mathbf{P} - \mathbf{P}\begin{pmatrix} \mathbf{0} \\ \mathbf{I} \end{pmatrix}\begin{pmatrix} \mathbf{0} & \mathbf{I} \end{pmatrix}\mathbf{P} \tag{8.44}$$

subject to

$$\mathbf{P}(t_f) = \mathbf{S} .$$

The solution to these equations, i.e., the guidance law, is

$$\mathbf{u}(t) = \mathbf{C}(t)\mathbf{x}(t) = -\frac{3\tau}{\frac{3}{C_1} + \tau^3}\begin{pmatrix} 1 & 0 & \tau & 0 \\ 0 & 1 & 0 & \tau \end{pmatrix}\mathbf{x}(t) , \tag{8.45}$$

where $\mathbf{C}$, it is recalled, is the control gain matrix.

It is important to note that $\mathbf{u} = -\mathbf{a}_{M_c}$ by definition and that it has components along both the $x$ and $y$ axes of the IFOC. If we let the reference LOS be the $x$-axis of our IFOC , we can see that the results shown in the example of Sec. 8.3.1 follow directly from the present one: $x_2$ becomes $y$, $\mathbf{F}$ is a $2 \times 2$ matrix, $\mathbf{S}$ and $\mathbf{R}$ are the scalars $C_1$ and 1, respectively, and $u$ (also a scalar) is given by the equation

$$u = \mathbf{C}\mathbf{x} = -\frac{3\tau}{\frac{3}{C_1} + \tau^3}\begin{pmatrix} 1 & \tau \end{pmatrix}\begin{bmatrix} y \\ \dot{y} \end{bmatrix} ,$$

which is identical to (8.2).

(b) The second example deals with a generalized guidance problem, such that capturing T is not the sole goal: We present a *rendezvous* case, where it is desired that M should approach T at zero relative velocity [9]; in the terminology of this section, simultaneously to achieve zero $\|\mathbf{r}_f\|$ and zero $\|\mathbf{v}_f\|$.

With (planar) NCC geometry and nonmaneuvering target, the system state is the vector $\mathbf{x} = [y \ v]^T$ where $v = \dot{y}$, and the control $u$ is $-a_M$, such that the state equation is

$$\dot{\mathbf{x}} = \begin{bmatrix} \dot{y} \\ \dot{v} \end{bmatrix} = \begin{pmatrix} 0 & 1 \\ 0 & 0 \end{pmatrix}\begin{bmatrix} y \\ v \end{bmatrix} + \begin{bmatrix} 0 \\ 1 \end{bmatrix} u . \tag{8.46}$$

The matrices in the performance index $J$ are $\mathbf{R} = 1$ (a scalar) and

$$\mathbf{S} = \begin{pmatrix} B_1 & 0 \\ 0 & B_2 \end{pmatrix} ,$$

so that, finally, the optimal-control problem is to minimize $J$ given by the equation

$$J = \frac{1}{2}(B_1 y_f^2 + B_2 v_f^2) + \frac{1}{2}\int_{t_i}^{t_f} a_M^2 dt . \tag{8.47}$$

(Note that, for $B_2 = 0$, (8.47) is precisely the performance index (8.1)).

The solution to the problem has been shown in [9] to be

$$a_M(t) = \frac{\frac{\tau}{B_2} + \frac{\tau^2}{2}}{D(\tau)}y(t) + \frac{\frac{1}{B_1} + \frac{\tau^2}{B_2} + \frac{\tau^3}{3}}{D(\tau)}v(t) \tag{8.48}$$

where

$$D(\tau) \triangleq \left(\frac{1}{B_1} + \frac{\tau^3}{3}\right)\left(\frac{1}{B_2} + \tau\right) - \frac{\tau^4}{4} .$$

If both $B_1$ and $B_2 \to \infty$, i.e., perfect rendevous ($y_f = 0$, $v_f = 0$) must be attained at any control cost, the law derived from (8.48) is

$$a_M(t) = \frac{2}{\tau^2}[3y(t) + 2v(t)\tau] . \tag{8.49}$$

Since $y = v_C \lambda \tau$ and $\dot{y} = v_C(-\lambda + \tau \dot{\lambda})$, an alternative form of (8.49) is

$$a_M(t) = v_C \left( \frac{2\lambda}{\tau} + 4\dot{\lambda} \right) .$$

Comparing this result with the law (8.3) obtained in Sec. 8.3.1 with perfect *intercept* in mind rather than rendezvous, we see that in addition to a term proportional to the LOS angle rate, the rendezvous law has a term which is proportional to the angle itself.

## 8.5   Laws Based on LQG Theory

### 8.5.1   Background

The engagements examined in Sections 8.3 and 8.4 have been described in a *deterministic* formulation. The resulting OCG laws required knowing the exact full state of the system throughout the engagement. In practice, however, the individual state variables cannot in general be determined exactly by direct measurements. Rather, measurements are made of a small number of variables, e.g., range, range rate, LOS rates, and body accelerations, which are functions of the state variables. These measurements are often corrupted by random errors, or noise, and the system itself may be subject to random disturbances such as wind or, for that matter, target maneuver. Thus, a *state estimator* will usually have to be developed and a *stochastic* formulation[3] will turn out to be more realistic than the deterministic one.

In the stochastic formulation it is understood that it is not in general possible for the guidance system to achieve zero miss distance, as it is in the deterministic case. Rather, one wishes the guidance to attain sufficiently low values of *mean-square* miss distance at reasonably low values of (mean-square) control effort, expecting small rms misses to be associated with large rms control effort and vice versa.

The performance index for the planar NCC engagement of Sec. 8.3.1 now becomes

$$J = E\left\{ \frac{1}{2} C_1 y^2(t_f) + \frac{1}{2} \int_{t_i}^{t_f} a_M^2 dt \right\} \tag{8.50}$$

where $E\{.\}$ denotes the expected value. Since $y_{f_{rms}}$, the rms value of $y_f$, is $\sqrt{E(y_f{}^2)}$ if $y_f$ has zero mean, the performance index is proportional to $y_{f_{rms}}^2$ when the integral in (8.50), which represents control effort, is ignored.

---

[3] It is assumed in Sec. 8.5 that the reader is aquainted with basic concepts of probability theory and random processes.

## 8.5.2   The LQG Problem

We have seen that the OCG law based on the solution to the optimal-control LQ problem is a linear feedback of the form $\mathbf{u}(t) = \mathbf{C}(t)\mathbf{x}(t)$. If this law is combined with *optimal estimation*, or *optimal filtering*, an overall optimum solution is obtained, provided the noise signals are Gaussian. This fact results from the *separation theorem*. The relevant theory is called *linear quadratic gaussian* (LQG) [3,4,9,32]. For a literature review of stochastic control theory applied to guidance, see reference [8].

We will now examine an example where *process noise* and *measurement noise* are present.

(a) Suppose the state of the system is known perfectly, but the initial conditions $\mathbf{x}_i \triangleq \mathbf{x}(t_i)$ are random. Suppose also that a *process noise* $\mathbf{w}$, assumed to be zero-mean gaussian white, is present, such that the model of the system dynamics is

$$\dot{\mathbf{x}}(t) = \mathbf{F}(t)\mathbf{x}(t) + \mathbf{G}(t)\mathbf{u}(t) + \mathbf{w}(t) \tag{8.51}$$

with

$$E\{\mathbf{w}(t)\} = 0, \quad E\{\mathbf{w}(t)\mathbf{w}^T(t')\} = \mathbf{A}(t)\delta(t - t') \tag{8.52}$$

and

$$E\{\mathbf{x}_i\} = 0, \quad E\{\mathbf{x}_i \mathbf{x}_i^T\} = \mathbf{X}_i .$$

$E\{\mathbf{w}(t)\mathbf{w}^T(t')\}$ in (8.52) is the covariance matrix of the process noise $\mathbf{w}$, $\mathbf{A}$ is positive-semidefinite, and $\delta(.)$ is the Dirac delta function. $E\{\mathbf{x}_i \mathbf{x}_i^T\}$ is the covariance matrix of the initial conditions $\mathbf{x}_i$.

The performance index is similar to (8.34) of the LQ problem, except that $J$ is now an expected value:

$$J = E\left\{\frac{1}{2}\mathbf{x}_f^T\mathbf{S}\mathbf{x}_f + \frac{1}{2}\int_{t_i}^{t_f}[\mathbf{x}^T(t)\mathbf{Q}(t)\mathbf{x}(t) + \mathbf{u}^T(t)\mathbf{R}(t)\mathbf{u}(t)]\,dt\right\}. \tag{8.53}$$

Since it is impossible to predict $\mathbf{w}(t')$ for $t' > t$ even with perfect knowledge of the state for $t' \leq t$, the solution to the present (stochastic) problem turns out to be identical to the solution obtained for the deterministic one, namely (8.35), (8.36) [9]. However, it can be shown that the (average) value obtained for the performance index $J$ is higher than the value obtained in the LQ case, as could be expected: *the presence of process noise degrades (on the average) the quality of optimal-control guidance.*

(b) In order to implement optimal control, the state variable $\mathbf{x}$ must of course be known, i.e., calculated from measured variables. However, in practice the measurement is never ideal: *measurement noise* $\mathbf{v}$ is also present, such that rather than x one actually has the state vector z, where z is related to x by the *measurement equation*

$$z(\tau) = H(\tau)x(\tau) + v(\tau), \quad t_i \leq \tau < t,\tag{8.54}$$

v being a zero-mean gaussian white noise characterized by the equations

$$E\{v(t)\} = 0, \quad E\{v(t)v^T(t')\} = B(t)\delta(t - t') .\tag{8.55}$$

The models of the system and the process noise are the same as in (a).

The LQG problem is to find u as a functional of $z(\tau)$, $t_i \leq \tau \leq t$, such that the performance index (8.53) be minimized.

The solution is

$$u(t) = C(t)\hat{x}(t) ,$$

which is similar to the (8.39) except that the feedback is $\hat{x}$, the estimated value of x, rather than x itself. The estimate is found by solving the equation

$$\dot{\hat{x}} = F\hat{x} + Gu + L[z - H\hat{x}] , \quad \hat{x}_i = 0 .\tag{8.56}$$

Details on methods for finding L can be found in the references cited above.

This way of obtaining the estimate of x is optimal in a certain way and is called *optimal estimation*, or *Kalman filtering* [33]. Although the theory of optimal estimation is beyond the scope of this book, one may get a rough idea of how it works from (8.56) and the diagram shown in Fig. 8.4. In the Estimator part of the diagram, $\hat{F}$, $\hat{G}$, and $\hat{H}$ are used rather than $F$, $G$, and $H$ in order to stress the fact that these are mere models which may not be known precisely. The Systam part of the diagram, where $F$, $G$, and $H$ appear, is a linear model of the real system: the mechanized subsystems are the Controller and the Estimator.

The control gain matrix C is as in (8.38), i.e., the same as for the deterministic case, P being found by solving the Ricatti equation (8.36). Thus, the LQG OCG law is seen to treat the estimates of x as if they were exact and use them in the optimal controller.

The value of $J$ achieved (on the average) in this case is greater than in the previous one, when process noise only was present, as might be expected. Furthermore, *the (rms) miss distances cannot be made smaller than the (rms) errors involved in measuring the respective variables that determine them.* In Bryson's words, "you cannot control any better than you can measure" [4].

## 8.6   On the Mechanization of OCG Laws

In previous sections of this chapter, several difficulties involved with the mechanization of OCG laws have been alluded to. We shall briefly deal with them in the present section.

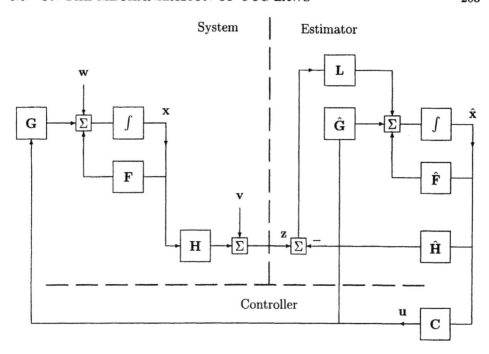

Figure 8.4: Flow diagram of optimal estimation and control

## 8.6.1 Control Acceleration

The optimal control $\mathbf{u}(t)$ for guiding M has been shown to require, in general, accelerations $\mathbf{a}_M(t)$ that have components in all three axes. There are mainly two difficulties here.

(a) With most present day missiles and similar guided objects, there is little, if any, control over the acceleration of M along its centerline axis, $a_{M_x}$. In practice one has control, albeit limited, on the lateral components of $\mathbf{a}_M$ only. Until variable-thrust engines are developed for this type of application, there seems to be no way out of this difficulty.

(b) Next we come to the constraints on the *magnitude* of $\mathbf{u}$ (we equate $\mathbf{u}$ with $-\mathbf{a}_{M_c}$). In practice, neither $a_{M_y}$ nor $a_{M_z}$ can exceed a certain value, say $a_{M_{max}}$, which in missiles and aircraft depends on altitude and speed. In Chapter 6 we have examined the detrimental effects of this limitation on the quality of PN guidance. In the PN case, one could reduce the requirements for $a_{M_{max}}$ by reducing the navigation constant, but this would be at the cost of increasing miss distances. In formulating OCG laws such a reduction can be achieved by choosing high weighting factors in the appropriate terms of the performance index; for example, by increasing

$C_2$ in (8.1) (which is equivalent to reducing $C_1$ in guidance laws (8.2), (8.7), etc.) or the appropriate elements of $\mathbf{R}$ in (8.34). In spite of this, $a_M$-saturation effects very often exist.

In OCG literature, the constraint on the magnitude of $\mathbf{u}$ is usually expressed by the inequality

$$\mathbf{u}^T(t)\,\mathbf{u}(t) \leq u_{max}^2 \tag{8.57}$$

where $u_{max} \equiv a_{M_{max}}$ [32]. The effects of a constraint expressed in this way on OCG laws for M with lag-free and single-lag guidance dynamics, respectively, have been studied by Anderson for planar NCC engagement conditions [22]. For example, for the case of lag-free dynamics with performance index (8.1), the OCG law

$$a_M(t) = \frac{3\tau}{\frac{3}{C_1} + \tau^3} m(t, \tau) \ , \tag{8.58}$$

derived in Sec. 8.3.1, changes into

$$a_M(t) = a_{M_{max}} sat\left(\frac{1}{a_{M_{max}}} \frac{3\tau}{\frac{3}{C_1} + \tau^3}\, m(t, \tau)\right) , \tag{8.59}$$

where the saturation function $sat(.)$ is as defined in (6.21). The OCG law (8.17) for single-lag dynamics changes in a similar way.

The effects of the constraint (8.57) on Rusnak and Meir's guidance laws for high-order dynamic have been shown to be quite similar [19, 20]. The general OCG law $a_{M_c}(t) = \Lambda(\tau)m(t, \tau)$ changes to

$$a_{M_c}(t) = a_{M_{max}} sat\left(\frac{1}{a_{M_{max}}}\Lambda(\tau)m(t, \tau)\right) . \tag{8.60}$$

## 8.6.2   Control Dynamics

In spite of the existence of OCG laws for dynamics of any order, the dynamics is very often assumed to be of first order, i.e., described by the transfer function

$$\frac{a_M}{a_{M_c}}(s) = \frac{1}{1 + sT_M} \ .$$

This approximation is justified by the fact that the dynamics of M is faster than the dynamics of the guidance process until a very short time before intercept; furthermore, the full-order OCG law does not generally result in improved performance with respect to first-order approximation (see Sec. 8.3.5). However, the approximation must be good, since the accuracy of the guidance depends on it strongly. For example, an error of 40% in the estimation of $T_M$ may lead to instability; smaller errors would increase miss distances to unacceptable levels without quite causing instability [34]. $T_M$ is either estimated and fixed before the guidance process begins or is constantly estimated during the process, and the guidance law utilized is (8.17)-(8.19) (or, equivalently, (8.26)-(8.29)).

Long experience with simulations has shown that if the representation of M's dynamics is not exact, the dynamics had better be fast in order to minimize miss distances. However, in practice there is an upper limit to the speed of response of the dynamics or, in other words, a lower limit to the equivalent time-constant of $G$ of any OCG system (or any PN guidance system, for that matter—see Sec. 6.3.5). Its value depends on input noises, radome aberration, and, finally, on physical properties of structure materials.

### 8.6.3 Radome Refraction Error

The effects of radome refraction (or aberration) error on PN guidance have been studied in Sec. 6.4.4; it has been noted there that, in radar missiles, this is one of the most important contributors to miss distance. When optimal-control guidance is involved, the effects are even worse, particularly for negative slope-coefficients $R$ (this term is elaborated on in Sec. 6.4.4). An example is shown in Fig. 6.15(b), the graph marked "OCG": it is seen that negative $\rho = kR/v_M$ of even 1.25% cannot be tolerated.

Some of the ways to reduce the effects of radome refraction error on the accuracy of PN guidance apply as well to optimal-control guidance; see Sec. 6.4.4.

### 8.6.4 Estimating the Time-to-Go

All OCG laws are formulated in terms of time-to-go $\tau$, and it is well known that accurate estimation of $\tau$ is crucial for the good performance of optimal-control guidance. In many approximative analyses, $\tau$ is assumed to equal $-r/\dot{r}$. However, this may be a good estimate only for small values of $\tau$ such that $v_C(\tau)$ is practically constant.

Assuming that the *estimated time-to-go* $\hat{\tau}$ is related to $\tau$ according to the equation $\hat{\tau} = A\tau + B$, the effects of errors in the estimation were studied for a certain OCG case. It was found that negative-$B$ errors led to instability, and that when total estimation error exceeded about 35%, PN achieved smaller miss distances than OCG [34].

Several algorithms have been proposed for estimating $\tau$ during the guidance process. Since detailed description of them is beyond the scope of this text, the reader may consult references [28, 35-41]. We note, however, that for all of the algorithms, measurement or estimation of range $r$ and closing velocity $v_C$ is absolutely necessary. Whereas such measurements are straightforward in active-homing radar systems, this is not the case in passive seekers, e.g., IR or TV. Such seekers do not directly measure either $r$ or $v_C$; rather, they measure angles ('bearings') and angle rates. Yet, 'Bearings Only' means for estimating $\tau$ in such systems have been studied as early as in 1968 [41] and seem to be developing fast.

## 8.6.5   Estimating the System State

As seen in Sec. 8.5.1, OCG laws based on LQG theory are combinations of optimal feedback control and optimal state estimation. Optimal estimators, or *Kalman filters*, are therefore ever-present components of OCG systems, used for estimating the system state and, indirectly, the time-to-go [33].

Target acceleration $\mathbf{a}_T$ is one of the state variables; naturally, estimating it presents great difficulties, especially when T executes evasive maneuvers and angle-only measurements are used by M [8,42,43]. Errors in estimating $\mathbf{a}_T$ are less detrimental to the accuracy of OCG systems with fast dynamics than to systems with slower ones. Note that a similar remark has been made in Sec. 8.6.2 regarding the representation of M's dynamics in OCG laws.

We will now present two examples.

(a) For a first example, a Kalman filter for estimating $y$, $\dot{y}$ and $a_T$ in an OCG system is shown, formulated in terms of a planar, NCC engagement [34]. The measurement input $y_{in} = \hat{r}\lambda$ is corrupted by noise caused by two effects, namely random target maneuver and glint noise, described by their spectral densities $\Phi_{tm}$ and $\Phi_g$, respectively. The filter can be represented by the transfer function

$$\frac{\hat{y}}{y_{in}}(s) = \frac{1 + 2\frac{s}{\omega_0} + 2(\frac{s}{\omega_0})^2}{1 + 2\frac{s}{\omega_0} + 2(\frac{s}{\omega_0})^2 + (\frac{s}{\omega_0})^3} \tag{8.61}$$

where $\hat{y}$ is the estimated value of $y$, and $\omega_0$ is given by

$$\omega_0 = \left(\frac{\Phi_{tm}}{\Phi_g}\right)^{\frac{1}{6}}. \tag{8.62}$$

(It may be interesting to compare this equation with (6.45) which was derived for PN). The structure of the filter, or estimator, is shown in Fig. 8.5, which also shows the way the filter interacts with the guidance-control part of the system. The inputs to the filter are $y_{in}$ and $a_M$, and the output of the guidance-control unit is $a_{M_c}$ [9, 34]. The coefficients $A_i, i = 1, ..., 4$ are functions of $\hat{\tau}$ and $T_M$ (a single-lag OCG law is implemented).

(b) More details are shown in the second example, where Kalman filters are used for estimating $a_T$, $\lambda$, $v_C$, and $\tau$ in order to implement a single-lag OCG law of the form (8.19) (see Fig. 8.6). It follows that the coefficients $A_3(\hat{\tau})$ and $A_4(\hat{\tau})$ of the previous example are now $f(\hat{\tau}/T_T)N'(\hat{\tau}/T_M)$ and $-f(\hat{\tau}/T_M)N'(\hat{\tau}/T_M)$, respectively, and that the sum $A_1(\hat{\tau})\hat{y} + A_2(\hat{\tau})\hat{\dot{y}}$ is replaced by $N'(\hat{\tau}/T_M)\hat{v}_C\lambda$ [44].

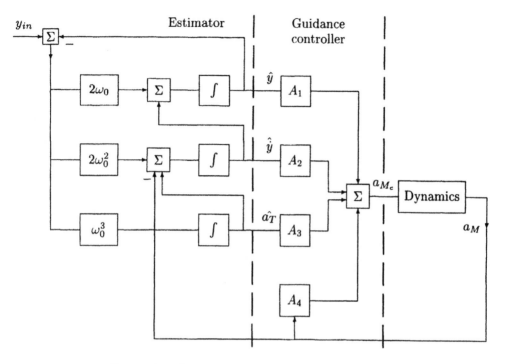

Figure 8.5: Optimal estimator in an OCG system

The filters have the following inputs:

- (Electronic) measurements of $r$ and $\dot{r}$,

- (Inertial instrumentation) measurements of $a_{M_y}$ and $a_{M_z}$,

- LOS inertial rates measured by rate gyroscopes, $\omega_y$ and $\omega_z$,

- LOS error-angles $e_y$ and $e_z$ (see e.g. Figs. 6.11-6.12),

- Information on the process noise statistics.

They provide the estimates $\hat{r}$, $\dot{\hat{r}}$, $\hat{\lambda}_y$, $\hat{\lambda}_z$, $\hat{a}_{M_y}$ and $\hat{a}_{M_z}$, as well as $\hat{a}_{T_y}$ and $\hat{a}_{T_z}$. From these estimates, $\hat{\tau}$ is obtained and, finally, the guidance law. In the diagram, Fig. 8.6, only one channel is depicted.

In this implementation, the time constants $T_M$ and $T_T$ are shown as known and fixed a priori: in practice, the dynamics of M and T, as represented by those time-constants, would rather be constantly estimated during the guidance process.

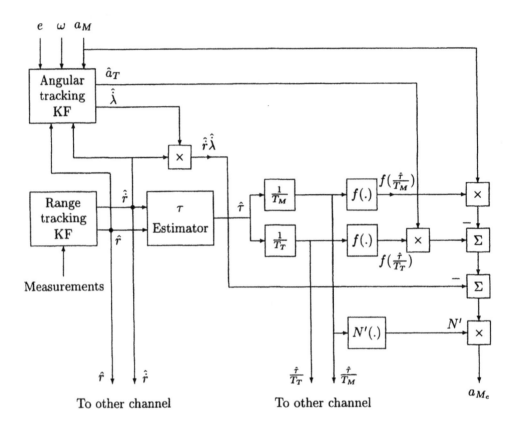

Figure 8.6: Implemention of a single-lag OCG law

## 8.7   Comparison with Other Guidance Laws

### 8.7.1   OCG and Proportional Navigation

It has been pointed out earlier in this chapter that, unlike the classical guidance laws, OCG laws cannot be said to be algorithms for directly implementing a certain geometrical rule; rather, they originate from the mathematical principle of optimization which is applicable to guidance as well as to a great many other systems. Nevertheless, they are not an exception of course to the requirement for reference to a line of sight.

OCG laws have been shown to have some similarities to PN; since both OCG and PN laws are relevant for modern air-to-air and surface-to-air missiles, a comparison between the two laws, indeed families of laws, is called for.

OCG laws seem to overcome the two main deficiencies of PN, namely sensitivity to noise and target maneuver, respectively. OCG requires smaller lateral accelerations $a_M$ than PN in order to capture a maneuvering target, which is an extremely important advantage from the engineering, hence economical, point of view. For example, in a comparison study, OCG required an $a_M/a_T$ acceleration advantage of 2.3, whereas PN required 5 [34].

On the other hand, OCG laws are less robust than PN ones, being more sensitive to radome refraction errors and being sensitive to errors in estimating the time-to-go and the system state and in modeling the system dynamics . Furthermore, optimal-control guidance generally requires more measurement instrumentation and more computing capability, which implies higher costs for the guidance unit, although the costs of relevant computer 'hardware' keep decreasing.

Capture zones, or launch envelopes, of OCG-guided missiles are generally larger than those of PN-guided ones. This is another very important advantage of the former over the latter. For example, planar launch envelopes of a hypothetical radar air-to-air missile are shown in Fig. 8.7 for four guidance laws, as follows.

(a) Classical PN, $a_{M_c} = N'v_C\dot{\lambda}_S$; $\lambda_S$ filtered by a first-order $G_F(s)$ (see Sec. 6.2).

(b) Same as (a), but improved estimation of $\dot{\lambda}$ (see Note (iii) at the end of Sec. 6.4.3).

(c) Zero-order OCG law (8.16); $f(\xi)$ approximated by the expression $f(\xi) \approx (3 + \xi)/(6 + 4\xi + \xi^2)$ (a Padé approximation).

(d) First-order OCG law (8.19); $f(\tau/T_T)$ approximated as in (c) and $f(\tau/T_M)$ approximated by a higher-order Padé approximation.

The envelopes have been obtained by simulating engagements such that T starts a high-$g$ maneuver sometime in the last 5 seconds before intercept. The maneuver start-moment and the maneuver direction are random. Since the launch envelopes are symmetric about the $v_T$ axis, only a half of each envelope is depicted. The diagrams are drawn to the same scale.

It is interesting to note that a very significant improvement is achieved in this example by changing from classical PN (a) to improved one (b), whereas the advantage of first-order OCG (d) over zero-order one (c) and improved PN (b) is not outstanding.

The improvement in capture-zones achieved by OCG compared to PN is shown to be unimpressive in another study, where the outer boundaries of the launch envelopes are compared [45]. On the other hand, great improvements in the inner boundaries have been shown by Riggs and by Dowdle et al., respectively [28,31].

## 8.7.2 OCG and Other Modern Laws

The OCG laws described in Sections 8.1-8.6 belong to a general class which can be called *Linear Quadratic Guidance Laws*. To this class belong laws based on optimal

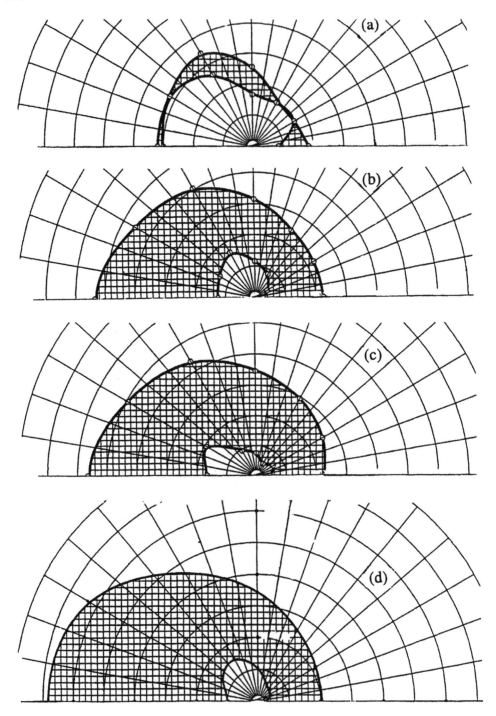

Figure 8.7: Launch envelopes for a hypothetical air-to-air missile

control theory and laws based on the theory of *differential games*.

From the differential-games point of view, M and T participate in a 'game' in which T is maneuvering in order to maximize the miss distance and M's control effort while minimizing its own control effort, and M is doing the opposite. In contrast to optimal control theory, the differential-games approach makes no assumption on *future* target maneuvers. Rather, it takes into consideration the target's maneuver *capabilities*. "The guidance law then guides M so as to minimize the potential effects of T's intelligent use of his maneuver capabilities" [46]. Although applying differential-games theory to problems of guidance is not new (see, e.g., references [1, 46-49]), it seems that this approach has not yet been applied in practical systems. This is one of the reasons we consider the topic to be beyond the scope of this text.

## 8.8 References

[1] Ho, Y.C., A.E. Bryson Jr., and S. Baron, "Differential Games and Optimal Pursuit-Evasion Strategies", *IEEE Tran. Aut. Cont.*, Vol. AC-10, No. 4, 1965, pp. 385-389.

[2] Bryson Jr., A. E., "Linear Feedback Solutions for Minimum Effort Interception, Rendezvous, and Soft Landing", *AIAA J.*, Vol. 3, No. 8, 1965, pp. 1542-1544.

[3] Kishi, F. H. and T. S. Bettwy, "Optimal and Sub-optimal Designs of Proportional Navigation Systems", in Lavi, A. and T. P. Vogl (eds.), *Recent Advances in Optimization Techniques*, John Wiley, 1965, pp. 519-540.

[4] Bryson Jr., A. E., "Aplications of Optimal Control Theory in Aerospace Engineering", *Minta Martin Lecture*, M.I.T., 1966; *J. Spacecraft*, Vol. 4, No. 5, 1967, pp. 545-553.

[5] Gonzalez, J., "New Methods in Terminal Guidance and Control of Tactical Missiles", Paper No. 3 in Maney, C. T. (ed.), "Guidance and Control for Tactical Guided Weapons with Emphasis on Simulation and Testing", *AGARD Lecture Series* No. 101, 1979, AD-A071129.

[6] Pastrick, H. L., S. M. Seltzer, and M. E. Warren, "Guidance Laws for Short-Range Tactical Missiles", *J. Guidance*, Vol. 4, No. 2, 1981, pp.98-108.

[7] Kelly, William C., "Homing Missile Guidance - A Survey of Classical and Modern Techniques", *Proc. IEEE Southcon*, Atlanta, 1981.

[8] Cloutier, J. R., J. H. Evers, and J. J. Feeley, "An Assessment of Air-to-Air Missile Guidance and Control Technology", *Proc. Amer. Contr. Conf.*, 1988, pp. 133-142.

[9] Bryson, A. E. and Yu-Chi Ho, *Applied Optimal Control*, Waltham, Mass., Blaisdell, 1969.

[10] Kreindler, E., "Optimality of Proportional Navigation", *AIAA J.*, Vol. 11, No. 6, 1973, pp. 878-880.

[11] Garber, V., "Optimum Intercept Laws for Accelerating Targets", *AIAA J.*, Vol. 6, No. 11, 1968, pp. 2196-2198.

[12] Willems, G., "Optimal Controllers for Homing Missiles", *U. S. Army Missile Command*, Report No. RE-TR-68-15, 1968.

[13] Asher, R. B. and J. P. Matuszewski, "Optimal guidance with Maneuvering Targets", *J. Spacecraft*, Vol. 11, No. 3, 1974, pp. 204-206.

[14] Forte, I. and J. Shinar, "Can a Mixed Guidance Strategy Improve Missile Performance?", *J. Guidance*, Vol. 11, No. 1, 1988, pp. 53-59.

[15] Chadwick, W. R. and P. Zarchan, "Interception of Spiraling Ballistic Missiles", *Proc. Amer. Cont. Conf.*, 1995, pp. 4476-4483.

[16] Cottrell, R. G., "Optimal Intercept Guidance for Short-Range Tactical Missiles", *AIAA J.*, Vol. 9, No. 7, 1971, pp. 1414-1415.

[17] Willems, G. C., "Optimal Controllers for Homing Missiles with Two Time Constants", *U. S. Army Missile Command*, Report No. RE-TR-69-20, 1969.

[18] Stockum, L.A. and F.C. Weimer, "Optimal and Suboptimal Guidance for a Short-Range Homing Missile", *IEEE Trans. Aero. Elect. Syst.*, Vol. AES-12, No. 3, 1976, pp. 355-360.

[19] Rusnak, Ilan, and Levi Meir, "Optimal Guidance for Acceleration Constrained Missile and Maneuvering Target", *IEEE Trans. Aero. Elect. Syst.*, Vol. 26, No. 4, 1990, pp. 618-624.

[20] Rusnak, Ilan, and Levi Meir, "Optimal Guidance for High-Order and Acceleration Constrained Missile", *J. Guidance*, Vol. 14, No. 3, 1991, pp. 589-596.

[21] Rusnak, Ilan, and Levi Meir, "Modern Guidance Law for High-Order Autopilot", *J. Guidance*, Vol. 14, No. 5, 1991, pp. 1056-1058.

[22] Anderson, Gerald M., "Effects of Performance Index/Constraint Combinations on Optimal Guidance Laws for Air-to-Air Missiles", *Proc. NAECON*, 1979, pp. 765-771.

[23] Hammond, J.K., "The Optimality of Proportional Navigation", *University of Southhampton*, AASU Report No. 313, 1972.

[24] Mirande, M., M. Lemoine, and E. Dorey, "Application of Modern Control Theory to the Guidance of an Air-to-Air Dogfight Missile", paper No. 21 in Williams, O.C.(ed.), *Guidance and Control Aspects of Tactical Air-Launched Missiles*, AGARD Conf. Proc. 292, 1980, AD-A092606.

[25] Guelman, M., and J. Shinar, "Optimal Guidance Law in the Plane", *J. Guidance*, Vol. 7, No. 4, 1984, pp. 471-476.

[26] Idan, M., O. M. Golan, and M. Guelman, "Optimal Planar Interception with Terminal Constraints, *J. Guidance*, Vol. 18, No. 6, 1995, pp. 1273-1279.

[27] Gonzalez, Jesse M., "New Methods in the Terminal Guidance and Control of Tactical Missiles", *Proc. NAECON*, 1979, pp. 350-361. A fuller version, under the same title, in *AGARD Lecture Series No. 101*, 1979.

[28] Riggs, Tom L., "Linear Optimal Guidance for Short Range Air-to-Air Missiles", *Proc. NAECON*, 1979, pp. 757-764.

[29] Kim, M. and K. V. Grider, "Terminal Guidance for Impact Attitude Angle Constrained Flight Trajectories", *IEEE Trans. Aero. Elect. Syst.*, Vol. AES-9, No. 6, 1973, pp. 852-859.

[30] York, R. J. and H. L. Pastrick, "Optimal Terminal Guidance with Constraints at Final Time", *J. Spacecraft*, Vol. 14, No. 6, 1977, pp. 381-383.

[31] Dowdle, J. R., M. Athans, S. W. Gully, and A. S. Willsky, "An Optimal Control and Estimation Algorithm for Missile Endgame Guidance", *Proc. IEEE Decision and Control Conf.*, 1982, pp. 1128-1132.

[32] Deyst, J. J. and C. F. Price, "Optimal Stochastic Guidance Laws for Tactical Missiles", J. Spacecraft, Vol. 10, No. 5, 1973, pp. 301-308.

[33] Gelb, A. (ed.), *Applied Optimal Estimation*, MIT Press, 1974.

[34] Nesline, F. W. and P. Zarchan, "A New Look at Classical vs Modern Homing Missile Guidance", *J. Guidance*, Vol. 4, No. 1, 1981, pp. 78-85.

[35] Mehra, R. K. and R. D. Ehrich, "Air-to-Air Missile Guidance for Strapdown Seekers", *Proc. IEEE Decision and Control Conf.*, 1984, pp. 1109-1115.

[36] Riggs, Tom L., "Optimal Control and Estimation for Terminal Guidance of Tactical Missiles", paper No. 23 in Williams, O.C.(ed.), *Guidance and Control Aspects of Tactical Air-Launched Missiles*, AGARD Conf. Proc. 292, 1980, AD-A092606.

[37] Riggs, Tom L., "Estimating Time-to-Go for Use in Advanced Guidance Laws", *Proc. 3rd Meeting, Coordinating Group on Modern Control Theory*, 1981, AD-A109124, pp. 177-199.

[38] Vergez, Paul L., "Linear Optimal Guidance for an AIM-9L Missile", *J. Guidance*, Vol. 4, No. 6, 1981, pp. 662-663.

[39] Lee, G. K. F., "Estimation of the Time-to-Go Parameter for Air-to-Air Missiles", *J. Guidance*, Vol. 8, No. 2, 1985, pp. 262-266.

[40] Hull, D.G., J.J. Radke, and R.E. Mack, "Time-to-Go Prediction for Homing Missiles Based on Minimum-Time Intercepts", *J. Guidance*, Vol. 14, No. 5, 1991, pp. 865-871.

[41] Rawling, A. G., "Passive Determination of Homing Time", *AIAA J.*, Vol. 6, No. 8, 1968, pp. 1604-1606.

[42] Speyer, J. L., K. D. Kim, and Minjea Tahk, "Passive Homing Missile Guidance Law Based on New Target Maneuver Models", *J. Guidance*, Vol. 13, No. 5, 1990, pp. 803-812.

[43] Hepner, S. A. R. and H. P. Geering, "Adaptive Two-Time-Scale Tracking Filter for Target Acceleration Estimation", *J. Guidance*, Vol. 14, No. 3, 1991, pp. 581-588.

[44] Meir, Levi, "On the Terminal Guidance of Radar Missiles", *RAFAEL* Lecture Preprints, 1994.

[45] Cheng, V. H. L. and N. K. Gupta, "Advanced Midcourse Guidance for Air-to-Air Missiles", *J. Guidance*, Vol. 9, No. 2, 1986, pp. 135-142.

[46] Anderson, G. M., "Comparison of Optimal Control and Differential Game Intercept Missile Guidance Laws", *J. Guidance*, Vol. 4, No. 2, 1981, pp. 109-115.

[47] Gutman, S. and G. Leitman, "Optimal Strategies in the Neighbourhood of a Collision Course", *AIAA J.*, Vol. 14, No. 9, 1976, pp. 1210-1212.

[48] Gutman, S. "On Optimal Guidance for Homing Missiles", *J. Guidance*, Vol. 2, No. 4, 1979, pp. 296-300.

[49] Shinar, J. and S. Gutman, "Three-Dimensional Optimal Pursuit and Evasion with Bounded Control", *IEEE Trans. Auto. Cont.*, Vol. AC-25, No. 3, 1980, pp. 492-496.

## FURTHER READING

Isaacs, R., *Differential Games*, John Wiley, 1965.

Bryson, A. E. and Yu-Chi Ho, *Applied Optimal Control*, Waltham, Mass., Blaisdell, 1969.

Anderson, B. D. O. and J. B. Moore, *Linear Optimal Control*, Prentice-Hall, 1971.

Kwakernaak, H. and R. Sivan, *Linear Optimal Control Systems*, John Wiley, 1972.

Gelb, A. (ed.), *Applied Optimal Estimation*, MIT Press, 1974.

Grimble, M. J. and M. A. Johnson, *"Optimal Control and Stochastic Estimation,* Wiley-Interscience, 1988.

Lin, C. F., *Modern Navigation, Guidance, and Control Processing*, Prentice-Hall, 1991.

Lewis, F. L. and V. L. Syrmos, *Optimal Control*, 2nd ed., Wiley-Interscience, 1995.

# Appendix A

# Equations of Motion

In this appendix, some properties and formulas related to general, three-dimensional kinematics of guidance will be presented briefly.

## A.1    General

The guided object M, the target T, and the origin of an inertial frame of coordinates (inertial FOC, or IFOC) are located at the points $M$, $T$, and 0, respectively. The vectors $0M$ and $0T$ are denoted by $\mathbf{r}_M$ and $\mathbf{r}_T$, and their time derivatives are the velocities $\mathbf{v}_M$ and $\mathbf{v}_T$, respectively (Fig. A.1). Another differentiation provides the accelerations $\mathbf{a}_M$ and $\mathbf{a}_T$. The *range* $\mathbf{r} \triangleq MT$ equals the difference $\mathbf{r}_T - \mathbf{r}_M$, so that

$$\begin{cases} \mathbf{r} = \mathbf{r}_T - \mathbf{r}_M \\ \dot{\mathbf{r}} = \mathbf{v}_T - \mathbf{v}_M \\ \ddot{\mathbf{r}} = \mathbf{a}_T - \mathbf{a}_M \ . \end{cases}$$

By definition, $\mathbf{r}$ is along the *Line of Sight* (LOS), which is the ray that starts at $M$ and passes through $T$. $\dot{\mathbf{r}}$ is called the *relative velocity*; it can be resolved into two components in the following way:

$$\dot{\mathbf{r}} = \dot{r}\mathbf{1}_r + r\dot{\mathbf{1}}_r = \dot{r}\mathbf{1}_r + \mathbf{w} \times \mathbf{r} \ , \tag{A.1}$$

where $\mathbf{1}_r$ is the unit vector along $\mathbf{r}$, i.e., $\mathbf{1}_r = \mathbf{r}/r$, $\mathbf{w}$ is the rate of rotation of $\mathbf{r}$, and "$\times$" denotes vector product. Note that $\dot{r}\mathbf{1}_r$ is the component of $\dot{\mathbf{r}}$ parallel to $\mathbf{r}$ and $\mathbf{w} \times \mathbf{r}$ is the component of $\dot{\mathbf{r}}$ perpendicular to $\mathbf{r}$.

By multiplying (A.1) by $\mathbf{r}$ (a vector product), one obtains the equation

$$\mathbf{r} \times \dot{\mathbf{r}} = \dot{r}\mathbf{r} \times \mathbf{1}_r + \mathbf{r} \times (\mathbf{w} \times \mathbf{r}) \ .$$

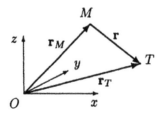

Figure A.1: Geometry of engagement

The first term on the right hand side of this equation obviously equals zero, and the second one equals $r^2 \mathbf{w}$ since $\mathbf{w}$ is perpendicular to $\mathbf{r}$.

(We note in passing that, for any vector $\mathbf{c}$ and any unit vector $\mathbf{1}_r$, the vector triple product $\mathbf{1}_r \times (\mathbf{c} \times \mathbf{1}_r)$ equals the component of $\mathbf{c}$ perpendicular to, or *across*, $\mathbf{1}_r$. We denote this component by $\mathbf{c}_\perp$. Also, the component of $\mathbf{c}$ *along* $\mathbf{1}_r$, denoted by $\mathbf{c}_\parallel$, is given by the product $\mathbf{1}_r(\mathbf{c} \bullet \mathbf{1}_r)$, where "$\bullet$" denotes scalar product. In other words,

$$\mathbf{c} = \mathbf{c}_\perp + \mathbf{c}_\parallel = \mathbf{1}_r \times (\mathbf{c} \times \mathbf{1}_r) + \mathbf{1}_r(\mathbf{c} \bullet \mathbf{1}_r) , \qquad (A.2)$$

respectively.)

Returning to the algebra preceding the note we see that we have

$$\mathbf{w} = \frac{\mathbf{r} \times \dot{\mathbf{r}}}{r^2} = \frac{\mathbf{r} \times (\mathbf{v}_T - \mathbf{v}_M)}{r^2} . \qquad (A.3)$$

The *closing velocity* $\mathbf{v}_C$, a term often used in guidance literature, is

$$\mathbf{v}_C \overset{\Delta}{=} -\dot{\mathbf{r}} = \mathbf{v}_M - \mathbf{v}_T . \qquad (A.4)$$

By multiplying (A.1) by $\mathbf{1}_r$ (a scalar product) we obtain the equation

$$\dot{\mathbf{r}} \bullet \mathbf{1}_r = (\dot{r}\mathbf{1}_r + \mathbf{w} \times \mathbf{r}) \bullet \mathbf{1}_r = \dot{r} \qquad (A.5)$$

since $\mathbf{1}_r \bullet \mathbf{1}_r = 1$ and $(\mathbf{w} \times \mathbf{r}) \bullet \mathbf{1}_r = 0$. Hence

$$\dot{r} = \frac{\mathbf{r} \bullet \dot{\mathbf{r}}}{r} = \frac{\mathbf{r} \bullet (\mathbf{v}_T - \mathbf{v}_M)}{r} . \qquad (A.6)$$

(Note that $\dot{r} \neq \|\dot{\mathbf{r}}\|$ unless $\dot{\mathbf{r}}$ and $\mathbf{r}$ are colinear).

For example, if $\mathbf{v}_T = [100\ 200\ 0]^T$, $\mathbf{v}_M = [0\ 400\ 600]^T$ and $\mathbf{r} = [2000\ 0\ 0]^T$ (in $m/sec$ and $m$, as appropriate), then, by (A.3),

$$\mathbf{w} = \frac{1}{2000^2} \begin{vmatrix} \mathbf{1}_x & 2000 & 100 - 0 \\ \mathbf{1}_y & 0 & 200 - 400 \\ \mathbf{1}_z & 0 & 0 - 600 \end{vmatrix} = \begin{bmatrix} 0 \\ 0.3 \\ -0.1 \end{bmatrix} \ rad/sec$$

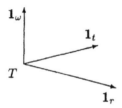

Figure A.2: A FOC centered at T

and, by (A.6),

$$\dot{r} = \frac{1}{2000} \begin{bmatrix} 2000 & 0 & 0 \end{bmatrix} \begin{bmatrix} 100 - 0 \\ 200 - 400 \\ 0 - 600 \end{bmatrix} = 100\,m/sec.$$

## A.2   A Rotating FOC

In some analyses, a rotating FOC whose axes are along the unit vectors $\mathbf{1}_w$, $\mathbf{1}_r$, and $\mathbf{1}_t$ is useful, where $\mathbf{1}_w = \mathbf{w}/\omega$, $\omega = \|\mathbf{w}\|$, and $\mathbf{1}_t = \mathbf{1}_w \times \mathbf{1}_r$ (Fig. A.2). Let the vector angular velocity of the rotating axes be $\mathbf{w}_{tri}$ and let $\mathbf{w}_{\|}$ denote the component of $\mathbf{w}_{tri}$ *along* $\mathbf{r}$, such that

$$\mathbf{w}_{tri} = \omega \mathbf{1}_w + \omega_{\|} \mathbf{1}_r , \tag{A.7}$$

where $\omega_{\|} = \|\mathbf{w}_{\|}\|$. It follows from this definition and the properties of vector products that one has

$$\dot{\mathbf{1}}_r = \mathbf{w}_{tri} \times \mathbf{1}_r = \omega \mathbf{1}_t$$

$$\dot{\mathbf{1}}_t = \mathbf{w}_{tri} \times \mathbf{1}_t = -\omega \mathbf{1}_r + \omega_L \mathbf{1}_w$$

$$\dot{\mathbf{1}}_w = \mathbf{w}_{tri} \times \mathbf{1}_w = -\omega_{\|} \mathbf{1}_t .$$

Using these equations one obtains the following results from (A.1) and (A.7).

$$\dot{\mathbf{r}} = \dot{r}\mathbf{1}_r + r\omega \mathbf{1}_t \tag{A.8}$$

$$\ddot{\mathbf{r}} = (\ddot{r} - r\omega^2)\mathbf{1}_r + (r\dot{\omega} + 2\dot{r}\omega)\mathbf{1}_t + r\omega\omega_{\|}\mathbf{1}_w . \tag{A.9}$$

The first term in (A.8) is often referred to as the *radial velocity*, and the second one as the *tangential velocity*.

   *Note.* In some kinematic problems, only one object, say M, is involved. In such cases we substitute $\mathbf{r}_M \equiv OM$ for $\mathbf{r}$ in equations (A.1)-(A.9).

## A.3   Coplanar Vectors

$\mathbf{v}_T$, $\mathbf{v}_M$ and $\mathbf{r}$ are said to be *coplanar* if they are on the same plane. A necessary and sufficient condition for this is that any of the *scalar triple products*

$$
\begin{cases}
\mathbf{v}_T \bullet (\mathbf{r} \times \mathbf{v}_M)\,, \\[1mm]
\mathbf{r} \bullet (\mathbf{v}_M \times \mathbf{v}_T)\,, \\[1mm]
\quad\text{or} \\[1mm]
\mathbf{v}_M \bullet (\mathbf{v}_T \times \mathbf{r})
\end{cases}
\tag{A.10}
$$

equal zero.[1] The three products are equivalent to each other.

Expressed in terms of the components of the vectors, this condition can be expressed as

$$
D \triangleq
\begin{vmatrix}
v_{T_x} & v_{T_y} & v_{T_z} \\
r_x & r_y & r_z \\
v_{M_x} & v_{M_y} & v_{M_z}
\end{vmatrix}
= 0\,.
\tag{A.11}
$$

Note that if any two of the three vectors are colinear, then $D = 0$.

Suppose $\mathbf{v}_T$, $\mathbf{v}_M$ and $\mathbf{r}$ are coplanar. Now in any engagement,

$$
\mathbf{v}_T \bullet \mathbf{w} = \mathbf{v}_T \bullet \frac{\mathbf{r} \times (\mathbf{v}_T - \mathbf{v}_M)}{r^2} = \frac{\mathbf{v}_T \bullet (\mathbf{r} \times \mathbf{v}_T) - \mathbf{v}_T \bullet (\mathbf{r} \times \mathbf{v}_M)}{r^2}
$$

(by (A.3)). The first term in the numerator equals zero due to properties of scalar and vector products; the second term equals zero in planar engagements, by (A.10). Therefore we have the equation

$$
\mathbf{v}_T \bullet \mathbf{w} = 0\,.
\tag{A.12}
$$

Similarly, $\mathbf{v}_M \bullet \mathbf{w} = 0$, and we already know that, by definition, $\mathbf{r} \bullet \mathbf{w} = 0$. Thus, in planar engagements, $\mathbf{w}$ is perpendicular to each of the vectors $\mathbf{v}_T$, $\mathbf{v}_M$ and $\mathbf{r}$, and colinear with both $\mathbf{r} \times \mathbf{v}_T$ and $\mathbf{r} \times \mathbf{v}_M$.

The angles between $\mathbf{r}$ and $\mathbf{v}_T$ and between $\mathbf{r}$ and $\mathbf{v}_M$, denoted by $\theta$ and $\delta$, are known as the *aspect angle* and the *lead angle*, respectively. In the planar case, both are defined about axes parallel to $\mathbf{w}$ (Fig. A.3). Therefore, by (A.3) and the definition of the vector product,

$$
w = \frac{v_T \sin\theta - v_M \sin\delta}{r}\,.
\tag{A.13}
$$

Similarly, (A.6) yields

---

[1] A geometrical interpretation of the scalar triple product $D = \mathbf{a} \bullet (\mathbf{b} \times \mathbf{c})$ is that $|D|$ equals the volume of a parallelepiped with the edges $a$, $b$, and $c$.

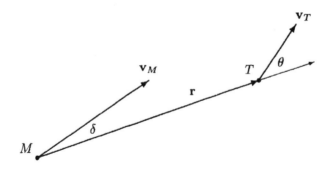

Figure A.3: Aspect angle $\theta$ and lead angle $\delta$

$$\dot{r} = v_T \cos\theta - v_M \cos\delta . \tag{A.14}$$

In contrast to (A.13), (A.14) is valid for nonplanar engagements as well.

$\theta$ can be found from the equation

$$\cos\theta = \frac{\mathbf{r} \bullet \mathbf{v}_T}{r v_T} . \tag{A.15}$$

It is defined to be positive or negative (we lose no generality by assuming $0 \le |\theta| \le \pi$) according to whether $\mathbf{r} \times \mathbf{v}_T$ is in the sense of $\mathbf{w}$ or opposite to it, respectively. In other words,

$$sgn(\theta) = sgn\{(\mathbf{r} \times \mathbf{v}_T) \bullet \mathbf{w}\} . \tag{A.16}$$

$\delta$ is found similarly, substituting $\mathbf{v}_M$ for $\mathbf{v}_T$ in (A.15) and (A.16). (Note: It follows from (A.16) and (A.3) that $v_M = 0 \Rightarrow \theta > 0$. Similarly, $v_T = 0 \Rightarrow \delta < 0$.)

*Note.* If $\mathbf{r}$ is colinear with $\mathbf{v}_T$, the engagement is obviously planar, as $D = 0$, and $\theta$ is either 0 or $180^o$; and similarly for $\mathbf{r}$, $\mathbf{v}_M$, and $\delta$, respectively. For the same reason, the engagement is also planar if $\mathbf{v}_T$ and $\mathbf{v}_M$ are colinear; in this case, either $\delta = \theta$ or $\delta = \theta \pm \pi$.

The case where all three vectors are colinear is trivial.

## A.4  Examples

**(a)** In the first example we deal with constant vectors. Let $\mathbf{v}_T, \mathbf{v}_M$ and $\mathbf{r}$ respectively be $\mathbf{v}_T = [-200 \ -100 \ 100]^T, \mathbf{v}_M = [-100 \ 400 \ 500]^T$, and $\mathbf{r} = [-3000 \ 6000 \ 9000]^T$ (in $m/sec$ and $m$, as appropriate). This is a planar engagement, since

$$D = \begin{vmatrix} -200 & -100 & 100 \\ -3000 & 6000 & 9000 \\ -100 & 400 & 500 \end{vmatrix} = 0 .$$

Since $\dot{\mathbf{r}} = [-100 \ -500 \ -400]^T$, we obtain the following results by (A.3) and (A.6).

$$\mathbf{w} = \frac{1}{r^2} \begin{vmatrix} \mathbf{1}_x & -3000 & -100 \\ \mathbf{1}_y & 6000 & -500 \\ \mathbf{1}_z & 9000 & -400 \end{vmatrix} = \frac{1}{60} \begin{bmatrix} 1 \\ -1 \\ 1 \end{bmatrix} \; rad/sec$$

and

$$\dot{r} = \frac{10^2 \times 10^3}{r} \begin{bmatrix} -3 & 6 & 9 \end{bmatrix} \begin{bmatrix} -1 \\ -5 \\ -4 \end{bmatrix} = -561 \, m/sec \, ,$$

where $r = \|\mathbf{r}\| = \sqrt{(-3000)^2 + 6000^2 + 9000^2} = 1000\sqrt{126}$ .
    Using (A.15) we find

$$\cos\theta = \frac{1}{r v_T} \begin{bmatrix} -200 & -100 & 100 \end{bmatrix} \begin{bmatrix} -3000 \\ 6000 \\ 9000 \end{bmatrix} \, ,$$

$$\cos\delta = \frac{1}{r v_M} \begin{bmatrix} -100 & 400 & 500 \end{bmatrix} \begin{bmatrix} -3000 \\ 6000 \\ 9000 \end{bmatrix} \, ,$$

where $v_T = \|\mathbf{v}_T\| = 100\sqrt{6}$ and $v_M = \|\mathbf{v}_M\| = 100\sqrt{42}$ . Hence

$$|\theta| = 70.9^\circ, \quad |\delta| = 8.2^\circ.$$

To determine the signs of the respective angles, we note that

$$\mathbf{r} \times \mathbf{v}_T = \begin{vmatrix} \mathbf{1}_x & -3000 & -200 \\ \mathbf{1}_y & 6000 & -100 \\ \mathbf{1}_z & 9000 & 100 \end{vmatrix} = 10^5 \begin{bmatrix} 15 \\ -15 \\ 15 \end{bmatrix} \, ,$$

$$\mathbf{r} \times \mathbf{v}_M = \begin{vmatrix} \mathbf{1}_x & -3000 & -100 \\ \mathbf{1}_y & 6000 & 400 \\ \mathbf{1}_z & 9000 & 500 \end{vmatrix} = 10^5 \begin{bmatrix} -6 \\ 6 \\ -6 \end{bmatrix} \, .$$

Since $\mathbf{w} = (1/60)[1 \ -1 \ 1]^T$, it follows, finally, by (A.16) that $\theta$ is positive and $\delta$ negative, i.e.,

$$\theta = 70.9^\circ, \quad \delta = -8.2^\circ.$$

For a check we can recalculate $w$ and $\dot{r}$, using (A.13) and (A.14), respectively.

$$w = \frac{100\sqrt{6} \sin 70.9^\circ + 100\sqrt{42} \sin 8.2^\circ}{1000\sqrt{126}} = 0.0289 \, \frac{rad}{sec} \, ,$$

which equals $(1/60)\sqrt{3}$, as obtained before, and

$$\dot{r} = 100\sqrt{6}\cos 70.9° - 100\sqrt{42}\cos 8.2° = -561\,m/sec.$$

(b) In the second example, we let the vectors be time-dependant. Suppose M is staionary at the origin of an inertial FOC while T climbs on the envelope of a straight cylinder along and around the $z$-axis. In other words, it describes a circular helix. The range $\mathbf{r}$ and the velocity $\mathbf{v}$ are given by the equations

$$\mathbf{r}(t) = \begin{bmatrix} x(t) \\ y(t) \\ z(t) \end{bmatrix} = \begin{bmatrix} c\cos\Omega t \\ c\sin\Omega t \\ vt\sin\gamma \end{bmatrix}, \quad \mathbf{v}(t) = \begin{bmatrix} -c\Omega\sin\Omega t \\ c\Omega\cos\Omega t \\ v\sin\gamma \end{bmatrix} \tag{A.17}$$

where $c$ is the radius of the cylinder, $\Omega$ is the angular frequency, $v$ is the speed of T, and $\gamma$ is the path angle. From (A.17) and the geometry of the case there follow the equations

$$c\Omega = v\cos\gamma \tag{A.18}$$

and

$$c = r\cos\theta, \quad z = r\sin\theta, \tag{A.19}$$

where $\theta$ is the elevation angle of T w.r.t. the $z = 0$ plane.

By (A.3) and (A.17)-(A.19) we obtain the following expression for $\mathbf{w}$:

$$\mathbf{w} = \frac{v}{r}\begin{bmatrix} \cos\theta(t)\sin\gamma\sin(\Omega t) - \sin\theta(t)\cos\gamma\cos(\Omega t) \\ -\cos\theta(t)\sin\gamma\cos(\Omega t) - \sin\theta(t)\cos\gamma\sin(\Omega t) \\ \cos\theta(t)\cos\gamma \end{bmatrix}. \tag{A.20}$$

(Recall that $r$ is a function of time). The acceleration $\mathbf{a}$, or $\dot{\mathbf{v}}$, is clearly

$$\mathbf{a} = \begin{bmatrix} -\Omega^2 c\cos(\Omega t) \\ -\Omega^2 c\sin(\Omega t) \\ 0 \end{bmatrix}. \tag{A.21}$$

Note that $a = \Omega^2 c$, i.e., the magnitude of $\mathbf{a}$ is constant and independant of $\gamma$. Applying (A.6) to (A.17) we get

$$\dot{r} = \frac{1}{r}(v\sin\gamma)^2 t. \tag{A.22}$$

$\ddot{r}$ is obtained by differentiating (A.22).

In order to get more insight into the geometry of the case, we now 'freeze' the kinematics at a point in time such that $\cos(\Omega t) = 1$ and $\sin(\Omega t) = 0$. The vectors in (A.17) and (A.20) become

$$\mathbf{r}(t) = \begin{bmatrix} c \\ 0 \\ z \end{bmatrix} = r\begin{bmatrix} \cos\theta \\ 0 \\ \sin\theta \end{bmatrix}, \quad \mathbf{v}(t) = v\begin{bmatrix} 0 \\ \cos\gamma \\ \sin\gamma \end{bmatrix} \tag{A.23}$$

and

$$\mathbf{w} = \frac{v}{r} \begin{bmatrix} -\sin\theta\cos\gamma \\ -\cos\theta\sin\gamma \\ \cos\theta\cos\gamma \end{bmatrix} . \qquad (A.24)$$

Since the rotation velocity vector of the FOC is

$$\mathbf{w}_{tri} = v \begin{bmatrix} 0 \\ -(c\sin\gamma)/r^2) \\ (\cos\gamma)/c \end{bmatrix} = \frac{v}{r} \begin{bmatrix} 0 \\ -\cos\theta\sin\gamma \\ (\cos\gamma)/\cos\theta \end{bmatrix} , \qquad (A.25)$$

$\mathbf{w}_{\parallel}$ is now easily obtained, being the difference $\mathbf{w}_{tri} - \mathbf{w}$. It is given by the equation

$$\mathbf{w}_{\parallel} = \frac{v\cos\gamma\tan\theta}{r} \begin{bmatrix} \cos\theta \\ 0 \\ \sin\theta \end{bmatrix} . \qquad (A.26)$$

Having found $\mathbf{r}$ and $\mathbf{w}$, hence $\mathbf{1}_r$ and $\mathbf{1}_w$, we obtain $\mathbf{1}_t$ by the product $\mathbf{1}_t = \mathbf{1}_w \times \mathbf{1}_r$.

## REFERENCES

Korn, G. A. and T. M. Korn, *Mathematical Handbook for Scientists and Engineers*, McGraw-Hill, 1968.

Regan, F. J. and S. M. Anandakrishnan, *Dynamics of Atmospheric Re-Entry*, Washington, D.C., AIAA, 1993.

# Appendix B

# Angular Transformations

Suppose $A$ and $B$ are three-dimensinal frames of coordinates (FOC), each having the three orthogonal axes $x$, $y$, and $z$, and $\mathbf{r}$ is a vector given in the $A$ FOC by its components $x$, $y$, and $z$. We wish to find the components of $\mathbf{r}$ in the $B$ FOC, which is rotated in a certain way with respect to $A$. This is a problem very often encountered in guidance work. For example, $A$ and $B$ may be the FOC's of an aircraft and inertial space, respectively; or $\mathbf{r}$ may be the position of a target T in M's seeker FOC and it is desired to know that position in M's body FOC.

Euler has found in the 18th century that every change in the relative orientation of $A$ and $B$ can be produced by rotation about a certain axis $L$ whose orientation relative to both $A$ and $B$ remains unaltered. In practice it is rather inconvenient to do calculations based on that axis. Instead, the rotation is done in three steps (in general), changing from the $(x, y, z)$ FOC to $(x', y', z')$, then to $(x'', y'', z'')$ and, finally, to $(x''', y''', z''')$, each rotation being done about one of the three axes. Such rotations are called *principal*.

It is quite straightforward to see that there are 12 distinct possible principal rotations; they are listed in Table B.1. These angular transformations are known as *Euler angle transformations*. Aeronautical and guidance engineers usually prefer transformations where all three axes are used, i.e., the even-numbered transformations of the table.

A $3 \times 3$ matrix is involved with each of the principal rotations. Let us denote the matrix by $\mathbf{T}$ with a subscript $x$, $y$, or $z$ according to the axis of the rotation. If the respective angles of rotation are $\phi$, $\theta$ and $\psi$ (always in the positive sense about the relevant axis), the transformation matrices are as follows.

$$\mathbf{T}_x(\phi) = \begin{pmatrix} 1 & 0 & 0 \\ 0 & \cos\phi & \sin\phi \\ 0 & -\sin\phi & \cos\phi \end{pmatrix} \tag{B.1}$$

Table B.1: The 12 principal rotations

| No. | 1st rotation, about | 2nd rotation, about | 3rd rotation, about |
|-----|---------------------|---------------------|---------------------|
| (1) | $x$ | $y'$ | $x''$ |
| (2) | $x$ | $y'$ | $z''$ |
| (3) | $x$ | $z'$ | $x''$ |
| (4) | $x$ | $z'$ | $y''$ |
| (5) | $y$ | $z'$ | $y''$ |
| (6) | $y$ | $z'$ | $x''$ |
| (7) | $y$ | $x'$ | $y''$ |
| (8) | $y$ | $x'$ | $z''$ |
| (9) | $z$ | $x'$ | $z''$ |
| (10) | $z$ | $x'$ | $y''$ |
| (11) | $z$ | $y'$ | $z''$ |
| (12) | $z$ | $y'$ | $x''$ |

$$\mathbf{T}_y(\theta) = \begin{pmatrix} \cos\theta & 0 & -\sin\theta \\ 0 & 1 & 0 \\ \sin\theta & 0 & \cos\theta \end{pmatrix} \tag{B.2}$$

$$\mathbf{T}_z(\psi) = \begin{pmatrix} \cos\psi & \sin\psi & 0 \\ -\sin\psi & \cos\psi & 0 \\ 0 & 0 & 1 \end{pmatrix}. \tag{B.3}$$

The angles in these matrices are called *Euler angles* (see also Fig. B̈.1, where $\odot$ denotes 'going out of the paper plane').

For a mnemonic aid note that, in each of the matrices (B.1)-(B.3),

(i) The main diagonal has only "1" and two cosines;

(ii) "1" is located in the 1st, 2nd or 3rd row (or column) according to whether the transformation is about $x$, $y$ or $z$, respectively;

(iii) $sin(.)$ has the $+$ sign in the row below the row with "1" (or, equivalently, the column preceding the column with "1").

Each of the matrices (B.1)-(B.3) has the following properties.

$$\mathbf{T}(-\alpha) = \mathbf{T}^T(\alpha), \ \ \mathbf{T}(\alpha+\beta) = \mathbf{T}(\alpha)\mathbf{T}(\beta) = \mathbf{T}(\beta)\mathbf{T}(\alpha), \ \ \mathbf{T}(0) = \mathbf{I}, \tag{B.4}$$

and

$$det(\mathbf{T}) = 1, \ \ \mathbf{T}\mathbf{T}^T = \mathbf{T}^T\mathbf{T} = \mathbf{I}, \ \ \mathbf{T}^{-1} = \mathbf{T}^T. \tag{B.5}$$

For any pair of the matrices (B.1)-(B.3), say $T_x(\phi)$ and $T_y(\theta)$, the product is not commutative, i.e., $T_x(\phi)T_y(\theta) \neq T_y(\theta)T_x(\phi)$ unless either of the angles is zero.

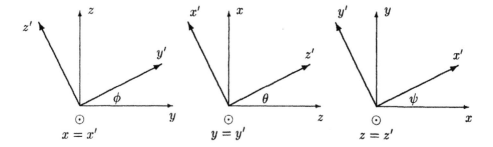

Figure B.1: The three principal rotations

However, the following property can easily be obtained from (B.4) and (B.5):

$$[T_x(\phi)T_y(\theta)]^{-1} = T_y(-\theta)T_x(-\phi) . \tag{B.6}$$

This property can be extended to any number of Euler matrices.

For example, suppose a vector **r** equals $[1\ 0\ 0]^T$ in $A$ FOC and we wish to express it in another FOC, say $B$ FOC, which is obtained by three successive rotations according to No. 12 of Table B.1., i.e.,

$$\mathbf{r}_B = \mathbf{T}_x(\phi)\mathbf{T}_y(\theta)\mathbf{T}_z(\psi)\mathbf{r}_A = \mathbf{T}_x(\phi)\mathbf{T}_y(\theta)\mathbf{T}_z(\psi)[1\ 0\ 0]^T. \tag{B.7}$$

The first rotation (about $z = z'$) gives

$$\begin{bmatrix} x' \\ y' \\ z' \end{bmatrix} = \begin{pmatrix} \cos\psi & \sin\psi & 0 \\ -\sin\psi & \cos\psi & 0 \\ 0 & 0 & 1 \end{pmatrix} \begin{bmatrix} 1 \\ 0 \\ 0 \end{bmatrix} = \begin{bmatrix} \cos\psi \\ -\sin\psi \\ 0 \end{bmatrix} .$$

The second rotation (about $y' = y''$) gives

$$\begin{bmatrix} x'' \\ y'' \\ z'' \end{bmatrix} = \begin{pmatrix} \cos\theta & 0 & -\sin\theta \\ 0 & 1 & 0 \\ \sin\theta & 0 & \cos\theta \end{pmatrix} \begin{bmatrix} \cos\psi \\ -\sin\psi \\ 0 \end{bmatrix} = \begin{bmatrix} \cos\theta\cos\psi \\ -\sin\psi \\ \sin\theta\cos\psi \end{bmatrix} .$$

$\mathbf{r}_B$ is finally obtained by a third rotation, about $x'' = x'''$, as follows.

$$\mathbf{r}_B = \begin{bmatrix} x''' \\ y''' \\ z''' \end{bmatrix} = \begin{pmatrix} 1 & 0 & 0 \\ 0 & c\phi & s\phi \\ 0 & -s\phi & c\phi \end{pmatrix} \begin{bmatrix} c\theta c\psi \\ -s\psi \\ s\theta c\psi \end{bmatrix} = \begin{bmatrix} c\theta c\psi \\ -c\phi s\psi + s\phi s\theta c\psi \\ s\phi s\psi + c\phi s\theta c\psi \end{bmatrix} ,$$

where $c(.) = \cos(.)$ and $s(.) = \sin(.)$ for the sake of brevity. For a check, one may verify that $x'''^2 + y'''^2 + z'''^2 = 1$.

We could, of course, skip the intermediate steps and obtain $\mathbf{r}_B$ by calculating $\mathbf{T}_{xyz}\mathbf{r}_A = \mathbf{T}_x\mathbf{T}_y\mathbf{T}_z\mathbf{r}_A$, where

$$
\mathbf{T}_{xyz} = \begin{pmatrix}
c\theta c\psi & c\theta s\psi & -s\theta \\
-c\phi s\psi + s\phi s\theta c\psi & c\phi c\psi + s\phi s\theta s\psi & s\phi c\theta \\
s\phi s\psi + c\phi s\theta c\psi & -s\phi c\psi + c\phi s\theta s\psi & c\phi c\theta
\end{pmatrix}. \tag{B.8}
$$

**Note.** The method for angular transformations presented in this Appendix is not the only one available. The method of *quaternions*, or the *four parameters of Euler*, may often be very useful. The reader is referred to the references for details.

## REFERENCES

Kaplan, Marshall H., *Modern Spacecraft Dynamics and Control*, John Wiley, 1976.

Regan, F. J. and S. M. Anandakrishnan, *Dynamics of Atmospheric Re-Entry*, Washington, D.C., AIAA, 1993.

Minkler, G. and J. Minkler, *Aerospace Coordinate Systems and Transformations*, Baltimore, Magellan Book Co., 1990.

American National Standard, *Atmospheric and Space Flight Vehicle Coordinate Systems*, ANSI/AIAA, R-004-1992, Feb. 1992.

# Appendix C

# A Few Concepts from Aerodynamics

## C.1   Skid-to-turn (STT) Configuration

Most present-day missiles have this configuration, also called *cruciform*, where a cylindrical body carries two sets of fixed wings and two sets of control surfaces. By convention, the axis along the centerline of the body is the $x$ axis, and the axes of the two sets of control surfaces, if they are $90^o$ apart, are the $y$ and $z$ axes, respectively. In this symmetric configuration, the wings are either in the $y = 0$ and $z = 0$ planes ("$+ +$" configuration) or rotated by $45^o$ about the $x$ axis ("$+\times$" configuration); see Figs. C.1(a) and (b), respectively.

Thus, if M has an STT configuration, then executing a turn in the $y = 0$ plane requires actuating control surfaces Nos. 1 and 3, and executing a turn in the $z = 0$ plane, control surfaces Nos. 2 and 4 (Fig. C.1). Angular motion of M about the $y$ axis is called *pitch*, and angular motion about $z$, *yaw*.

If either of the two sets of control surfaces (or both) is actuated *differentially*, there results a rotation of M about the $x$ axis, called *roll*. Sometimes the said control surfaces are used for pitch and yaw only, and a separate, third set of surfaces is used for roll control.

In most analytical studies of guidance it is assumed that the $y$ and $z$ control channels are decoupled from each other, i.e., that a command for yaw results in yaw motion only, and similarly for pitch. This of course is only an idealization, roughly valid for slow motions. No study is complete without taking coupling effects into account, which is usually done by simulation.

Roll motions are parasitic in STT control; in principle, no roll moments should exist at all. However, they do, due to the following reasons.

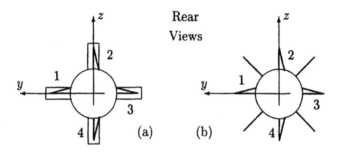

Figure C.1: (a) A "+ +" configuration, (b) A "+×" configuration

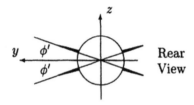

Figure C.2: Asymmetric STT configuration

(i) Asymmetries in structure,
(ii) Asymmetry of the thrust,
(iii) Atmospheric disturbances,
(iv) Interaction of the pitch and yaw channels.
Since it is impossible in practice to eliminate all of these perturbations, in particular the fourth one, roll control is almost always necessary.

In some guidance systems, the roll *angle* must remain constant, say zero, throughout the engagement; for example, in command-to-line-of-sight (CLOS) systems and in many TV-guided weapon systems. In other systems, the roll angle does not necessarily have to be zero, but the roll *rate* does, or must be kept sufficiently low.

*Note.* In a less symmetric STT configuration, the control surfaces axes (and the wings) form the angles (about the $x$ axis) $\pm\phi'$ and $180° \pm \phi'$ with the $y$ axis (Fig. C.2). ($\phi' = 45°$ in the symmetric case). In some configurations, the wings are mounted on a ring which is free to rotate about the $x$-axis.

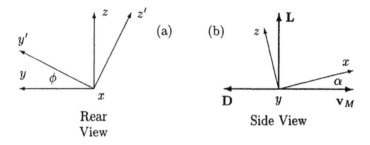

Figure C.3: (a) BTT control; (b) Lift, drag, and angle-of-attack

## C.2 Bank-to-turn (BTT) Configuration

This configuration, also called *Twist and Steer*, is familiar from conventional air-craft. Indeed, a few early guided weapons looked like small aircraft, e.g., the World War II German Hs-293 and the American GB Series. Interest in applying this configuration for modern tactical missiles is relatively recent, resulting from the fact that its asymmetric cross-section gives it large acceleration capability in its pitch plane. The well known Tomahawk cruise missile (not a tactical missile, of course, although it homes onto its target when it is sufficiently close to it) has this configuration.

Let us use the following notation. $x$ is the centerline forward axis of M, same as in STT configurations; the $y$ axis is along the wings, and $z$ completes a right-angle FOC. In order to execute a turn in a plane which is not its $y = 0$ plane, a BTT-controlled M must first bank, or twist, by a certain roll angle $\phi$ (Fig. C.3 (a)). This is due to the fact that the *lift* $L$ in this configuration is practically in the $y = 0$ plane only.[1] In STT configurations, on the other hand, the lift can in principle be in any plane, depending on the respective lifts created by the $y$ and $z$ channels. The amount of lift depends on the *angle of attack* $\alpha$, which is the angle (about the $y$ axis) between $\mathbf{v}_M$ and the $x$ axis of M (see Fig. C.3(b)).

## C.3 On Angle of Attack and Sideslip

Analogous to the angle of attack $\alpha$ which is defined about the $y$ axis there may exist a *sideslip* angle $\beta$ (about the $z$ axis). In BTT configurations, $\beta$ is kept very small and $\alpha$ positive. The latter constraint results from engine consideration. In STT configurations, there is, in principle, no distinction between $\alpha$ and $\beta$; the former relates to turning about the $y$ axis and the latter about the $z$ axis.

---

[1]It is convenient to regard the aerodynamic force that acts on M, in any configuration, as the vector sum of two forces, *lift* $\mathbf{L}$ and *drag* $\mathbf{D}$; the former is orthogonal to M's velocity $\mathbf{v}_M$ and the latter is opposite to it (Fig. C.3(b)).

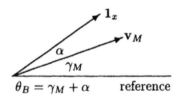

Figure C.4: Path angle $\gamma_M$, angle of attack $\alpha$, and attitude angle $\theta_B$

For a given configuration and a given flight condition (velocity, air density), the lift $L$ is roughly proportional to $\alpha$ (or $\beta$) when these angles are small, according to the equation

$$L = (\frac{1}{2}\rho v_M{}^2)CS\alpha,$$

where $(1/2)\rho v_M{}^2$ is the dynamic pressure, $\rho$ is the air density, $C$ is a coefficient, and $S$ is a reference area, usually the cross-section of the body. Since the lateral acceleration $a_M$ equals $L/m$, one has

$$a_M = K_{aer}\alpha, \tag{C.1}$$

where $K_{aer}$, the *aerodynamic gain*, is given by the equation

$$K_{aer} = \frac{\rho v_M{}^2 CS}{2m}. \tag{C.2}$$

Let us now introduce the *attitude angle*, or *body angle*, $\theta_B$ and the *path angle* $\gamma$, the angles about the $y$ axis that $\mathbf{1}_x$ and $\mathbf{v}_M$ form with an inertial reference line, respectively (Fig. C.4); $\mathbf{1}_x$ is the unit vector along the $x$-axis of M. In the plane defined by these vectors, $a_M = v_M\dot{\gamma}_M$ by kinematics, and

$$\theta_B = \gamma_M + \alpha \tag{C.3}$$

by the geometry. We therefore have

$$\dot{\theta}_B = \frac{a_M}{v_M} + \frac{\dot{a}_M}{K_{aer}} = \frac{1}{v_M}\left(1 + \frac{v_M}{K_{aer}}\frac{d}{dt}\right)a_M,$$

hence

$$\dot{\theta}_B = \frac{1}{v_M}\left(1 + T_\alpha\frac{d}{dt}\right)a_M, \tag{C.4}$$

where $T_\alpha \triangleq v_M/K_{aer}$. $T_\alpha$ is called the *turning rate time-constant* of M, since the turning rate $\dot{\gamma}_M$ is related to the angle of attack by the equation $\dot{\gamma}_M = \alpha/T_\alpha$. This time-constant is a measure of the maneuverability of M's airframe: the higher $T_\alpha$ is, the less maneuverable is M. It plays an important role in several guidance problems. By (C.2), its value is given by

$$T_\alpha = \frac{2m}{\rho v_M CS} .$$
(C.5)

Clearly, $T_\alpha$ is not really constant, being directly proportional to the mass $m$ and inversely proportional to the product $\rho v_M$. The coefficient $C$ changes with flight condition and, to a lesser extent, with $\alpha$ itself. Thus, $T_\alpha$ may attain high values when the missile is flying slowly at high altitudes, especially when it has not yet lost mass due to the burning of fuel.

## C.4   Note

When M maneuvers by *thrust vector control (TVC)* rather than by aerodynamic forces (for example, if it operates outside the atmosphere, or if its speed is too low for aerodynamic forces to be effective), the turning rate time-constant $T_\alpha$ is *defined* by the ratio $\alpha/\dot{\gamma}_M$. Assuming that the thrust vector $\mathbf{F}_{th}$ is directed roughly along the body longitudinal axis, it can easily be shown (see again Fig. C.4) that

$$a_M = \frac{F_{th} \sin \alpha}{m} .$$
(C.6)

Hence, since $\dot{\gamma}_M = a_M/v_M$, it follows that, provided $\alpha$ is small, $a_M \approx \frac{F_{th}\alpha}{m}$ and

$$T_\alpha = \frac{m v_M}{F_{th}} .$$
(C.7)

Thus, $T_\alpha$ can be regarded as the ratio between M's velocity and its axial acceleration.

### REFERENCES

Etkin, Bernard, *Dynamics of Flight - Stability and Control*, 2nd ed., John Wiley, 1982.

Mclean, Donald, *Automatic Flight Control Systems*, Prentice-Hall, 1990.

Blakelock, John, *Automatic Control of Aircraft and Missiles*, 2nd ed., Wiley-Interscience, 1991.

Jenkins, Philip N., "Missile Dynamic Equations for Guidance and Control Modeling and Analysis", *US Army Missile Command, Redstone Arsenal*, Technical Report RG-84-17, April 1984.

Arrow, A., "Status and Concerns for Bank-to-Turn Control of Tactical Missiles", *J. Guidance*, Vol. 8, No. 2, 1985, pp. 267-274.

# Appendix D

# Derivations of Several Equations

## D.1 The Graphs of the $Kh$ Plane, Sec. 2.3.2

From the definition of the engagement dealt with in Example 1, one has

$$\cot \lambda = \frac{x_T}{y_T} = \frac{x_{T_0} - v_T t}{h} , \tag{D.1}$$

hence

$$\frac{d}{dt}(\cot \lambda) = -\frac{v_T}{h} .$$

For the sake of brevity we now define

$$\xi \stackrel{\Delta}{=} \cot \lambda ,$$

so that $\dot{\xi} = -v_T/h$ by (D.1). Since

$$\frac{dx_M}{dt} = v_M \cos \gamma , \quad \frac{dy_M}{dt} = v_M \sin \gamma$$

(where $\gamma \equiv \gamma_M$ for brevity) by definition of $\gamma$ (see also Fig. 2.4), it follows that

$$\begin{cases} \frac{dx_M}{d\xi} = \frac{-v_M h \cos \gamma}{v_T} = -Kh \cos \gamma \\ \frac{dy_M}{d\xi} = \frac{-v_M h \sin \gamma}{v_T} = -Kh \sin \gamma . \end{cases} \tag{D.2}$$

Clearly, $x_M = \xi y_M$; differentiating this equation w.r.t. $\xi$ one gets

$$\frac{dx_M}{d\xi} = \xi \frac{dy_M}{d\xi} + y_M .$$

Hence, by (D.2),

$$-Kh\cos\gamma = -Kh\xi\sin\gamma + y_M \ . \tag{D.3}$$

By differentiating this equation w.r.t. $\xi$ and substituting $-Kh\sin\gamma$ for $dy_M/d\xi$ we get the equation

$$\sin\gamma\frac{d\gamma}{d\xi} = -2\sin\gamma - \xi\cos\gamma\frac{d\gamma}{d\xi}$$

which can be rearranged to have the form

$$(2\sin\gamma\frac{d\xi}{d\gamma} + \xi\cos\gamma) + \sin\gamma = 0 \ . \tag{D.4}$$

The sum in parentheses equals $2\sqrt{\sin\gamma}\frac{d}{d\gamma}(\xi\sqrt{\sin\gamma})$. Therefore (D.4) can be rewritten as

$$-2d(\xi\sqrt{\sin\gamma}) = \sqrt{\sin\gamma}d\gamma \ .$$

Integration now gives

$$2(\xi_0\sqrt{\sin\gamma_0} - \xi\sqrt{\sin\gamma}) = g(\gamma) - g(\gamma_0)$$

where

$$g(\gamma) \triangleq \int^\gamma \sqrt{\sin\gamma'}d\gamma' \ .$$

Recalling that $\xi = \cot\lambda$ by definition and $\gamma_0 = \lambda_0$ by (2.2)(ii), we finally get $\cot\lambda$ as a function of $\gamma$:

$$\cot\lambda(\gamma) = \frac{2\cot\lambda_0\sqrt{\sin\lambda_0} + g(\lambda_0) - g(\gamma)}{2\sqrt{\sin\gamma}} \ . \tag{D.5}$$

Using the path angle $\gamma$ as a parameter, the equations of the trajectories can now be written as follows

$$\begin{cases} y_M = Kh(\cot\lambda\sin\gamma - \cos\gamma) & \text{(by (D.3))} \\ x_M = y_M\cot\lambda \\ t = \frac{h}{v_T}(\cot\lambda - \cot\lambda_0) & \text{(by (D.1))} \end{cases}$$

where $\cot\lambda$ is given by (D.5); the function $g(.)$ in (D.5) can be expressed in terms of elliptic integrals, as follows [1, Eq. 2.595.2]:

$$g(\gamma) = -\sqrt{2}\left[2E\left(\alpha(\gamma), 1/\sqrt{2}\right) - F\left(\alpha(\gamma), 1/\sqrt{2}\right)\right] \ ,$$

where $F$ and $E$ are elliptic integrals of the first and second kind, respectively (with moduli $1/\sqrt{2}$), and $\alpha(.) \triangleq \arccos\sqrt{\sin(.)}$. It can also, more simply, be calculated numerically. Using (2.10), these equations are very easily transformed to nondimensional ones expressed in terms of the variables of the $Kh$ plane, as follows.

$$\begin{cases} y^* = (\cot\lambda\sin\gamma - \cos\gamma) \\ x^* = y^*\cot\lambda \\ t^* = \cot\lambda - \cot\lambda_0 \ . \end{cases}$$

## D.2   Derivation of (2.21)

Suppose that at a certain point in time, T is at the point $(c, 0, h)$ in an inertial $(x, y, z)$ FOC. It follows from the definition of the example that

$$\mathbf{r}_T = [c \ 0 \ h]^T, \quad \mathbf{v}_T = [0 \ v_T \ 0]^T, \quad \mathbf{v}_M = v_M[\cos \delta \sin \beta \quad \sin \delta \quad \cos \delta \cos \beta]^T,$$

and

$$\mathbf{w}_M = \mathbf{w}_T = \mathbf{w} = (1/r_T^2)\mathbf{r}_T \times \mathbf{v}_T = (v_T/r_T)[-\cos \beta \ 0 \ \sin \beta]^T .$$

We recall that $\delta = \arcsin(r_M/Kr_T)$ by (2.18) and $\beta = \arctan(c/h) = \arcsin(c/r_T)$.

We will now use the concept of the rotationg axes presented in Appendix A.2. By geometrical considerations we find that the angular rate of the rotating axes is $\mathbf{w}_{tri} = ((v_M/x) \sin \delta)[0 \ 0 \ 1]^T = (v_T/c)[0 \ 0 \ 1]^T$, hence

$$\mathbf{w}_{\parallel} = \mathbf{w}_{tri} - \mathbf{w} = \frac{v_T \cot \beta}{r_T}[\sin \beta \ 0 \ \cos \beta]^T.$$

By (A.9), $\mathbf{a}_M$ is given in the $\mathbf{1}_w, \mathbf{1}_r, \mathbf{1}_t$ rotating FOC by the equation

$$\mathbf{a}_M = r_M w w_{\parallel} \mathbf{1}_w + (\ddot{r}_M - r_M w^2)\mathbf{1}_r + (r_M \dot{w} + 2\dot{r}_M w)\mathbf{1}_t .$$

It follows from the geometry defined by vectors shown above that

$$w = \frac{v_T}{r_T} = \frac{v_T}{c} \sin \beta, \quad \dot{w} = 0, \quad w_{\parallel} = \frac{v_T}{c} \cos \beta . \tag{D.6}$$

$\dot{r}_M = v_M \cos \delta$ and, by (2.19),

$$\ddot{r}_M = -v_M \dot{\delta} \sin \delta = -\frac{v_M v_T \sin \delta \cos \beta}{h} .$$

By substitution into the expression for $\mathbf{a}_M$ and using the equation $a_T = v_T^2/c$, one obtains the equation

$$\frac{a_M}{a_T} = 2K \sqrt{\sin^2 \beta + (\frac{r_M \cos \beta}{2Kr_T})^2} ,$$

which is equivalent to (2.21).

## D.3   Proofs for (3.8) and (3.9)

We start by rewriting the kinematic equations (3.4).

$$\dot{r} = v_T \cos \theta - v_M \tag{D.7}$$

$$r\dot{\theta} = -v_T \sin\theta. \tag{D.8}$$

Multiplying (D.7) by $\cos\theta$ and (D.8) by $\sin\theta$ and substracting, one obtains the equation

$$\dot{r}\cos\theta + v_M \cos\theta - r\dot{\theta}\sin\theta = v_T.$$

Substituting (by (D.7)) $K(\dot{r} + v_M)$ for $v_M \cos\theta$, one gets

$$\frac{d}{dt}[r(K + \cos\theta)] = v_T(1 - K^2) \tag{D.9}$$

(it is recalled that $K = v_M/v_T$). Integrating (D.9), assuming $K \neq 1$, we obtain the expression (3.8) for $t$ in terms of $r$ and $\theta$, where $r_0$ and $\theta_0$ are the initial conditions:

$$t = \frac{r_0}{v_T} \frac{K + \cos\theta_0 - (\frac{r}{r_0})(K + \cos\theta)}{K^2 - 1} .$$

For $K = 1$, we recall the function $r(\theta)$, (3.7), in its general form, i.e.,

$$r(\theta) = r_0 \frac{\sin\theta_0}{\tan(\theta_0/2)} \frac{\tan(\theta/2)}{\sin\theta} . \tag{D.10}$$

Substituting (D.10) into (D.8) we get the differential equation

$$\frac{d\theta}{dt} = \frac{-v_T \sin^2\theta}{C_0 \tan\frac{\theta}{2}} ,$$

which can also be written in the form

$$\frac{dt}{d\theta} = \frac{-C_0}{v_T(1 + \cos\theta)\sin\theta} , \tag{D.11}$$

where

$$C_0 \triangleq \frac{r_0 \sin\theta_0}{\tan(\theta_0/2)} = r_0(1 + \cos\theta_0) .$$

Integrating (D.11), (3.9) is obtained:

$$t = \frac{r_0}{2v_T}\left[1 - \frac{1 + \cos\theta_0}{1 + \cos\theta} - (1 + \cos\theta_0)\log\frac{\tan\frac{\theta}{2}}{\tan\frac{\theta_0}{2}}\right] .$$

## D.4   On the $t_f$-Isochrones of Sec. 3.3.1(c)

Substituting $r = 0$ in (D.9) we see that a pursuit that starts at the point $(r_0, \theta_0)$ lasts

$$t_f = \frac{r_0}{v_T} \frac{K + \cos\theta_0}{K^2 - 1} , \quad K > 1 . \tag{D.12}$$

Note that $t_f$ depends only on the velocities $v_T$, $v_M$ and the initial conditions. We shall use (nondimensional) Cartesisn rather than polar coordinates, defining

$$x_0 \triangleq \frac{r_0 \cos \psi_0}{v_T t_f}, \quad y_0 \triangleq \frac{r_0 \sin \psi_0}{v_T t_f} \quad,$$

and $\psi_0 = -\theta_0$ by definition. (D.12) now becomes

$$K\sqrt{x_0^2 + y_0^2} + x_0 = K^2 - 1 .$$

From this equation, the familiar equation for an ellipse is obtained:

$$\left(\frac{x_0 + 1}{K}\right)^2 + \left(\frac{y_0}{K^2 - 1}\right)^2 = 1 .$$

This equation describes an ellipse centered at $(-1, 0)$ having the major semiaxis $K$ and the minor semiaxis $\sqrt{K^2 - 1}$, so that the origin $(0, 0)$ is one of its foci.

# D.5    Definition of DPP (Sec. 3.3.4) in Vector Terms

We first require that the vectors $\mathbf{v}_M$, $\mathbf{r}$, and $\mathbf{v}_T$ be coplanar. In mathematical terms, the equation

$$\mathbf{v}_M \bullet (\mathbf{r} \times \mathbf{v}_T) = 0 \tag{D.13}$$

must be satisfied (see Appendix A.3 for details on coplanarity). The fact that $\mathbf{v}_M$ leads $\mathbf{r}$ by the angle $\delta$ is expressed by the equation

$$\frac{\mathbf{v}_M \bullet \mathbf{r}}{v_M r} = \cos \delta . \tag{D.14}$$

($\cos \delta$ must be positive, otherwise one would have deviated pure escape rather than pursuit). Due to the coplanarity property, the vectors $\mathbf{r} \times \mathbf{v}_M$ and $\mathbf{r} \times \mathbf{v}_T$ are colinear; if $(\mathbf{r} \times \mathbf{v}_M) \bullet (\mathbf{r} \times \mathbf{v}_T) > 0$, or, equivalently, $\mathbf{v}_{M_\perp} \bullet \mathbf{v}_{T_\perp} > 0$, $\mathbf{v}_M$ is said to *lead* $\mathbf{r}$; in the opposite case it is said to *lag* behind it.

For a numerical illustration, suppose $\mathbf{r} = [-3000 \ 6000 \ 9000]^T$, $v_M = 500$, and $\mathbf{v}_T = [-200 \ -100 \ 100]^T$, (in $m$ and $m/sec$, as appropriate). One wishes to find the components of $\mathbf{v}_M$ necessary for having a lead of $27°$.

By substitution into (D.13) (see Apendix A.3 for details on scalar triple products) one obtains the equation

$$\begin{vmatrix} -200 & -100 & 100 \\ -3000 & 6000 & 9000 \\ v_{M_x} & v_{M_y} & v_{M_z} \end{vmatrix} = 0 ,$$

hence

$$v_{M_x} - v_{M_y} + v_{M_z} = 0 .$$

Substitution into (D.14) provides a second equation for the components, namely

$$1000(-3v_{M_x} + 6v_{M_y} + 9v_{M_z}) = 1000\sqrt{(-3)^2 + 6^2 + 9^2} \times 500 \cos 27° = 5 \times 10^6 .$$

The third equation necessary for solving our problem is

$$v_{M_x}^2 + v_{M_y}^2 + v_{M_z}^2 = v_M^2 = 25 \times 10^4 .$$

The three equations lead to a quadratic one, whose two solutions are

$$\mathbf{v}_{M_1} = [56.2 \ 378.3 \ 322.1]^T ,$$

$$\mathbf{v}_{M_2} = [-294.3 \ 97.9 \ 392.2]^T .$$

By examining the signs of the products $(\mathbf{r} \times \mathbf{v}_{M_1}) \bullet (\mathbf{r} \times \mathbf{v}_T)$ and $(\mathbf{r} \times \mathbf{v}_{M_2}) \bullet (\mathbf{r} \times \mathbf{v}_T)$ it is seen that $\mathbf{v}_{M_2}$ leads $\mathbf{r}$ whereas $\mathbf{v}_{M_1}$ lags behind it.

## D.6   A Proof for (4.11)

The squares of distances travelled by T and by M during the time-of-flight $t_f$ are

$$(v_T t_f)^2 = (x - \frac{r_0}{2})^2 + y^2$$

and

$$(v_M t_f)^2 = (x + \frac{r_0}{2})^2 + y^2,$$

respectively. Since $v_M t_f = K v_T t_f$, one gets the equation

$$(x + \frac{r_0}{2})^2 + y^2 = K^2[(x - \frac{r_0}{2})^2 + y^2] ,$$

from which (4.11) directly follows.

## D.7   A Proof for Inequality (4.13)

The equations that describe this engagement are (4.12), namely

$$\dot{r} = v_T \cos \theta - v_M \cos \delta \qquad\qquad (D.15)$$

$$v_M \sin \delta = v_T \sin \theta \qquad\qquad (D.16)$$

$$\dot{\theta} = \dot{\gamma}_T - \dot{\lambda} = \omega . \qquad\qquad (D.17)$$

Differentiating (D.16) w.r.t. time $t$ and substituting $\omega$ for $\dot{\theta}$ by (D.17) we obtain the equation

$$v_M \dot{\delta} \cos \delta = v_T \omega \cos \theta = a_T \cos \theta ,$$

since $v_T \omega = a_T$. Hence, since $\dot{\delta} = \dot{\gamma}_M$ in parallel navigation,

$$a_M = v_M \dot{\gamma}_M = a_T \frac{\cos \theta}{\cos \delta} .$$

Thus the acceleration ratio is

$$\left| \frac{a_M}{a_T} \right| = \left| \frac{\cos \theta}{\cos \delta} \right| = \frac{|\cos \theta|}{\sqrt{1 - \left( \frac{\sin \theta}{K} \right)^2}} . \tag{D.18}$$

Assuming of course $|(1/K) \sin \theta| < 1$, we conclude that

$$\left| \frac{a_M}{a_T} \right| \quad \begin{cases} \leq 1, & K > 1, \quad any \ \theta \\ = 1, & K = 1, \quad any \ \theta \\ = 1, & any \ K, \quad \theta = 0, \ 180^o \\ \geq 1, & K < 1, \quad any \ \theta , \end{cases} \tag{D.19}$$

which confirms (4.13).

# D.8  Derivation of (4.15) of Sec. 4.2.2(a)

From (D.15) (see above) and (4.14), which is

$$\int_0^{t_f} \dot{r}(t)dt = r(t_f) - r_0 = -r_0 ,$$

we get the equation

$$-r_0 = v_T \int_0^{t_f} (\cos \theta - K \cos \delta) \, dt = v_T \int_0^{t_f} [\cos \theta - K\sqrt{1 - \left( \frac{\sin \theta}{K} \right)^2} ] \, dt . \tag{D.20}$$

By definition of the engagement, $\theta(t) = \theta_0 + \omega t$ and $v_T/\omega = c$, where $c$ is the radius of T's circle. Therefore (D.20) can be rewritten in the form

$$-r_0 = c \int_{\theta_0}^{\theta_f} [\cos \theta - K\sqrt{1 - \left( \frac{\sin \theta}{K} \right)^2} ] \, d\theta .$$

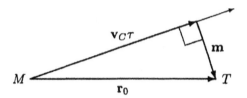

Figure D.1: Geometry of miss

From this integral it is straightforward to obtain the implicit equation (4.15) for $\theta_f$ and hence for $t_f = (\theta_f - \theta_0)/\omega$:

$$K \left[ E(\theta_f, \frac{1}{K}) - E(\theta_0, \frac{1}{K}) \right] - (\sin \theta_f - \sin \theta_0) = \frac{r_0}{c} , \quad K > 1 .$$

$E(., 1/K)$ is the elliptic integral of the second kind with the modulus $1/K$. It is also sometimes written as $E(., \alpha)$ where $\alpha \triangleq \arcsin(1/K)$ is the modular angle.

## D.9　Derivation Of (4.34) and (4.35)

Since both $\mathbf{v}_T$ and $\mathbf{v}_M$ are constant by assumption, M approaches T with a constant closing velocity $\mathbf{v}_C$. The vectors $\mathbf{r}_0$, $\mathbf{v}_C \tau$, and $\mathbf{m}$ form a right-angled triangle, where $\tau$ is the time until the closest approach and $\mathbf{m}$ is the miss distance (Fig. D.1). It can be shown (see e.g. [2]) that $\tau$ is given by the equation

$$\tau = \frac{\mathbf{r}_0 \bullet \mathbf{v}_C}{v_C^2} . \tag{D.21}$$

From the said right-angled triangle it follows that $m$ equals $r_0 \sin(\mathbf{r}_0, \mathbf{v}_C)$, which is (4.35).

$m$ can also be expressed in terms of the closing velocity, as follows. First we note the vector difference

$$\mathbf{m} = \mathbf{r}_0 - \mathbf{v}_C \tau = \mathbf{r}_0 - \mathbf{v}_C \frac{\mathbf{r}_0 \bullet \mathbf{v}_C}{v_C^2} . \tag{D.22}$$

We then express the vectors $\mathbf{r}_0$ and $\mathbf{v}_C$ by their components, i.e.

$$\mathbf{r}_0 = [r_0 \ 0]^T, \quad \mathbf{v}_C = [v_{C_x} \ v_{C_y}]^T .$$

Doing the scalar product it now follows from (D.22) that the components of the miss distance $\mathbf{m}$ are given by the equations

$$m_x = r_0(1 - \frac{v_{C_x}^2}{v_C^2}), \quad m_y = -r_0 \frac{v_{C_x} v_{C_y}}{v_C^2} ,$$

where $= [m_x \ m_y]^T = \mathbf{m}$. $m$ of course equals the square root of the sum $m_x^2 + m_y^2$, which is easily found to be $r_0 v_{C_y}/v_C$. Hence (4.34).

# D.10 Vector Representation for Sec. 5.4.1

**(a)** We first recall equation (A.9) for relative acceleration $\ddot{\mathbf{r}}$, namely

$$\ddot{\mathbf{r}} = (\ddot{r} - rw^2)\mathbf{1}_r + (r\dot{w} + 2\dot{r}w)\mathbf{1}_t + rww_\| \mathbf{1}_w = \mathbf{a}_T - \mathbf{a}_M . \qquad (D.23)$$

By guidance law (5.4) for TPN, $\mathbf{a}_{M_c} = k\mathbf{w} \times \mathbf{1}_r$. By definition of the axes $\mathbf{1}_w, \mathbf{1}_r, \mathbf{1}_t$, this guidance law implies that $\mathbf{a}_{M_c}$ is directed along the $\mathbf{1}_t$ axis. Also, by the ideal-dynamics assumption, $\mathbf{a}_M = \mathbf{a}_{M_c}$. Substitution into (D.23) yields the equation

$$r\dot{w} + 2\dot{r}w = \mathbf{a}_T \bullet \mathbf{1}_t - kw$$

(since $\mathbf{a}_T \bullet \mathbf{1}_t$ is the value of the component of $\mathbf{a}_T$ along the $\mathbf{1}_t$ axis). For the nonmaneuvering target case, this equation is equivalent to (5.14). For the two other axes, where no guidance control is exerted, one has

$$\ddot{r} - rw^2 = \mathbf{a}_T \bullet \mathbf{1}_r, \quad rww_\| = \mathbf{a}_T \bullet \mathbf{1}_w .$$

**(b)** We recall the vector difference of the previous section:

$$\mathbf{m} = \mathbf{r} - \mathbf{v}_C \tau$$

($\mathbf{r}$ is now used rather than $\mathbf{r}_0$; compare to (5.18) and (5.19)). Multiplying this equation by $\mathbf{v}_C/r^2$ we obtain

$$\frac{\mathbf{v}_C \times \mathbf{m}}{r^2} = \frac{\mathbf{v}_C \times \mathbf{r}}{r^2} .$$

By (A.3), the right-hand side of this equation equals $\mathbf{w}$, the angular rate of the LOS. Therefore one has

$$\mathbf{w} = \frac{\mathbf{v}_C \times \mathbf{m}}{r^2} .$$

Thus, the TPN law (5.4) can be expressed as

$$\mathbf{a}_{M_c} = k\frac{\mathbf{v}_C \times \mathbf{m}}{r^2} \times \mathbf{1}_r . \qquad (D.24)$$

Furthermore, if the angle between $\mathbf{v}_C$ and the LOS $\mathbf{r}$ is small, then (D.21) gives $\tau = r/v_C$; it then follows from (D.24) that the guidance law simplifies into

$$\mathbf{a}_{M_c} = k\frac{\mathbf{v}_C \times \mathbf{m}}{(v_C \tau)^2} \times \mathbf{1}_r = N'\frac{(\mathbf{v}_C/v_C) \times \mathbf{m}}{\tau^2} \times \mathbf{1}_r$$

where $N' = k/v_C$. This equation is the vector version of (5.20).

## D.11   On Equivalent Noise Bandwidth

The integral

$$\frac{1}{2|G(j\omega_0)|^2} \int_{-\infty}^{\infty} |G(j\omega)|^2 \, d\omega \,, \tag{D.25}$$

where $G(.)$ is the transfer function of a filter and $\omega_0$ its center frequency, is called the *equivalent noise bandwidth* of the filter, denoted by $\omega_{enb}$ [3]. For the filters examined in Chapter 6, $\omega_0 = 0$ and $G(0) = 1$. Hence, $\omega_{enb}$ for a single-lag $G(s)$ equals $\pi/2T_M$, since

$$\frac{1}{2} \int_{-\infty}^{\infty} \frac{d\omega}{1 + (\omega T_M)^2} = \frac{\pi}{2T_M} \,.$$

For a quadratic $G$, i.e.,

$$G(s) = \frac{1}{1 + 2\zeta\frac{s}{\omega_n} + (\frac{s}{\omega_n})^2} \,,$$

$\omega_{enb} = (\pi/4)(\omega_n/\zeta)$ [4].

## D.12   APN Law in Vector Terms

The *vector* zero-effort miss ditance **m** can be expressed by the series

$$\mathbf{m}(t,\tau) = \mathbf{r} + \dot{\mathbf{r}}\tau + \frac{\mathbf{a}_T}{2}\tau^2 + \frac{\dot{\mathbf{a}}_T}{6}\tau^3 + \ldots \,.$$

Ignoring terms with power higher than 2 and using the alternative definition of TPN (D.24), namely $\mathbf{a}_{M_c} = (k/r^2)(\mathbf{v}_C \times \mathbf{m}) \times \mathbf{1}_r$, one obtains the law as

$$\mathbf{a}_{M_c} = \frac{k}{r^2}(\mathbf{v}_C \times \mathbf{r} + 0 + \mathbf{v}_C \times \frac{\mathbf{a}_T}{2}\tau^2) \times \mathbf{1}_r \,,$$

which simplifies to

$$\mathbf{a}_{M_c} = (k\mathbf{w} + \frac{k}{r^2}\mathbf{v}_C \times \frac{\mathbf{a}_T}{2}\tau^2) \times \mathbf{1}_r \,.$$

Assuming now NCC conditions, such that the approximations $\tau = r/v_C$ and $\mathbf{1}_{v_C} = \mathbf{1}_r$ are valid, and recalling the definition $N' = k/v_C$, one finally has

$$\mathbf{a}_{M_c} = N'(v_C\mathbf{w} \times \mathbf{1}_r + \frac{\mathbf{a}_{T_\perp}}{2}) \,, \tag{D.26}$$

where $\mathbf{a}_{T_\perp}$ is the component of $\mathbf{a}_T$ perpendicular to $\mathbf{r}$, since $(\mathbf{1}_r \times \mathbf{a}_T) \times \mathbf{1}_r = \mathbf{a}_{T_\perp}$ (by (A.2)). This is the required APN law.

For an example, suppose $\mathbf{r}$ and $v_C$ are in the $z = 0$ plane, which we shall call the *engagement plane*, and suppose the angle between them is small, such that the approximations made in deriving (D.26) are valid. Let **r** coincide with the $x$ axis. Then clearly **w** has a $z$ component only, i.e., $\mathbf{w} = [0\ 0\ \omega]^T$. If T starts a maneuver $\mathbf{a}_T = [a_{T_x}\ a_{T_y}\ a_{T_z}]^T$, then, by (D.26), M's acceleration command $\mathbf{a}_{M_c}$ (assuming M is guided by APN) would be

$$\mathbf{a}_{M_c} = N' \begin{bmatrix} 0 \\ v_C\omega \\ o \end{bmatrix} + \frac{N'}{2} \begin{bmatrix} 0 \\ a_{T_y} \\ a_{T_z} \end{bmatrix} ,$$

since the component $a_{T_x}$ does not contribute to $\mathbf{a}_{T_\perp}$. Thus, it may be advantageous for T to concetrate its maneuvering capability in the $z$ direction; this will make M change the direction of its control acceleration which, prior to T's maneuver, had a $y$ component only. In other words, T may gain by maneuvering *out of the engagement plane*. This of course is true for nonaugmented PN as well.

## D.13  Derivation of (8.14)

By the assumptions, T's maneuver across the LOS is given by the equation

$$a_T(t) = A\sin(\omega t + \psi) ,$$

where the amplitude $A$, the frequency $\omega$, and the phase $\psi$ are constant. Hence, by (8.13),

$$m_T(t, \tau) = \int_0^\tau (\tau - t')A\sin[\omega(t + t') + \psi]\, dt'.$$

Calculating this integral one obtains

$$m_T(\tau) = \frac{A}{\omega^2}\{\omega\tau\cos(\omega t + \psi) + \sin(\omega t + \psi) - \sin[\omega(t + \tau) + \psi]\} . \tag{D.27}$$

By expressing the term $\sin[\omega(t + \tau) + \psi]$ as the sum $\sin(\omega t + \psi)\cos\omega\tau + \cos(\omega t + \psi)\sin\omega\tau$ and recalling that $a_T(t) = A\sin(\omega t + \psi)$ and $\dot{a}_T(t) = A\omega\cos(\omega t + \psi)$ we see that (D.27) is equivalent to (8.14).

## D.14  References

[1] Gradshteyn, I. S. and I. M. Ryzhik, *Table of Integrals, Series and Products*, Academic Press, 1965.

[2] Abzug, M. J., "Vector Methods in Homing Guidance", *J. Guidance*, Vol. 2, No. 3, 1979, pp. 253-255.

[3] Barton, D. K., *Modern Radar System Analysis*, Norwood, Mass., Artech House, 1988.

[4] Garnell, P., *Guided Weapon Control Systems*, 2nd ed., Pergamon Press, 1980.

# List of Symbols and Abbreviations

| | |
|---|---|
| $\mathbf{a}_M$ | Lateral acceleration of M (a vector) |
| $a_M$ | Lateral acceleration of M (a scalar) $(= \|\mathbf{a}_M\|)$ and similarly for other vector quantities |
| $a_T$ | Lateral acceleration of T |
| $c$ | A constant distance |
| $c(.)$ | $cos(.)$ |
| $D$ | A constant distance |
| $D(s)$ | Parasitic disturbance |
| $e$ | Error in a closed loop |
| $f$ | Frequency |
| $f(.)$ | Weighting function |
| $F(s), G(s), H(s)$ | Transfer functions |
| $g$ | Acceleration of Gravity |
| $h$ | A constant distance, often vertical |
| $J$ | Performance index |
| $K$ | Ratio between M and T velocities, i.e., $v_M/v_T$ |
| $k, k'$ | Gain constants |
| $m$ | Miss distance |
| $N$ | Proportional-Navigation constant |
| $N'$ | Effective Proportional-Navigation constant |
| $R$ | Refraction slope coefficient |
| $r$ | Range |
| $s$ | Laplace perator |
| $s(.)$ | $sin(.)$ |
| $t$ | Time |
| $t_f$ | Time of flight |
| $T_d, T_e, T_1, T_2$ | Time constants |
| $T_\alpha$ | Aerodynamic turning-rate time-constant |
| $T_M$ | (Equivalent) time-constant of M |
| $T_T$ | (Equivalent) time-constant of T |
| $v_M$ | Velocity of M |
| $v_T$ | Velocity of T |
| $v_C$ | Closing velocity = - range rate |
| $U$ | A constant velocity |

| | |
|---|---|
| $\mathbf{w}$ | Vector angular velocity |
| $x_M, y_M, z_M$ | Coordinates of M |
| $x_T, y_T, z_T$ | Coordinates of T |
| | |
| $\alpha$ | angle of attack |
| $\beta$ | An angle |
| $\beta$ | A positive constant not greater than 1 |
| $\gamma$ | Path angle |
| $\delta$ | Angle between $\mathbf{v}_M$ and line-of-sight |
| $\epsilon$ | Base of natural logarithms |
| $\varepsilon$ | A constant |
| $\varepsilon_{RR}$ | Error due to radome refraction |
| $\zeta$ | Damping coefficient in 2nd order system |
| $\theta$ | Angle between $\mathbf{v}_T$ and line-of-sight |
| $\theta_B$ | Attitude, or Body angle |
| $\lambda$ | Angle of line-of-sight relative to a fixed line |
| $\mu$ | Angle of deviated line-of-sight relative to a fixed line |
| $\xi$ | Nondimensional time-to-go |
| $\rho$ | Radius of curvature |
| $\rho$ | Refraction coefficient, proportional to $R$ |
| $\sigma$ | Intensity, or rms value |
| $\tau$ | Time to go |
| $\Phi$ | Power spectral density |
| $\psi$ | Aspect angle |
| $\psi_S$ | Angle of line-of-sight relative to M's axis |
| $\omega$ | Angular velocity ($= \|\mathbf{w}\|$) |
| $\omega$ | Angular frequency ($= 2\pi f$) |
| $\omega_b$ | Bandwidth |

Subscripts

| | |
|---|---|
| $c$ | Command |
| $f$ | Final |
| $fn$ | Fading noise |
| $g$ | Glint noise |
| $i$ | Initial |
| $M$ | Related to Missile M, or pursuer |
| $max$ | maximum or maximal |
| $min$ | minimum or minimal |
| $n$ | Noise |
| $o$ | Initial |
| $R$ | Refraction |
| $rbm$ | Right beam |

| | |
|---|---|
| $rms$ | Root-mean-square |
| $RR$ | Due to radome refraction |
| $S$ | Sight |
| $T$ | Related to target T, or evader |
| $W$ | Wind |
| $x$ | $x$ component |
| $y$ | $y$ component |
| $z$ | $z$ component |
| $\perp$ | Component across the line of sight |
| $\parallel$ | Component along the line of sight |

### Other Notations

| | |
|---|---|
| $()^*$ | Nondimensional |
| $\dot{()}$ | $d()/dt$ |
| $\hat{()}$ | Estimated value of () |
| $\bullet$ | Scalar product |
| $\times$ | Vector product |
| $\triangleq$ | Defined, or by definition |

### Abbreviations

| | |
|---|---|
| ACPN | Acceleration-compensated proportional navigation |
| APN | Augmented proportional navigation |
| BPN | Biased proportional navigation |
| BR | Beam rider, beam riding |
| BTT | Bank to turn |
| CAN | Constant aspect navigation |
| CC | Collision course |
| CLOS | Command to line-of-sight |
| DLG | Dynamic lead guidance |
| DPP | Deviated pure pursuit |
| EMD | Error measuring device |
| FOC | Frame of coordinates |
| FOV | Field of view |
| GPN | Generalized proportional navigation |
| IFOC | Inertial Frame of Coordinates |
| IP | Intercept point |
| IR | Infra-red |

| LOS | Line of sight |
| LOT | Linear optical trajectory |
| LQ | Linear Quadratic |
| LQG | Linear Quadratic Gaussian |
| M | Missile, or pursuer |
| NCC | Near collision course |
| OCG | Optimal-control law |
| PIP | Projected intercept point |
| PLG | Proportional lead guidance |
| PN | Proportional navigation |
| PP | Pure pursuit |
| PPN | Pure proportional navigation |
| RF | Radio frequency |
| rms | Root mean square |
| RTPN | Rate-using true proportional navigation |
| SO | Switch-over |
| STT | Skid to turn |
| T | Target, or evader |
| TGL | Terminal guidance law |
| TPN | True proportional navigation |
| TT | Target's tail |
| TV | Television |
| 3-D | three-dimensions, three-dimensional |

# Index

Acceleration, lateral, 2. See also kinematics
   definition of, 8
accuracy, xv, 24, 130, 172, 204-5
acoustic, 4, 43, 49
adjoint, 132, 160
Adler, F.P., 103
aerodynamic(s), 37, 70, 73, 86, 131, 149, 151, 196, 229-233
aircraft, xiii, 2, 3, 16, 48-49, 83, 85, 88, 94, 95, 158, 185
   pilotless, 12, 103
air defense, 6, 16, 41, 168, 182
air-to-air, 4, 14, 54, 146, 172, 182, 196, 208-209
air-to-ground, 175
air-to-sea, 5, 13, 176
air-to-surface, 30
Alpert, J., 158
altimeter, 173, 176
America(n), 231
Anderson, G.M., 204
angle of attack, 69, 117, 144, 166, 197, 231-232
angular transformation, 7, 28, 90, 225-228
ant(s), 59, 61
antenna, 154
anti-aircraft, 24
anti-ballistic, 182
anti-ship, 41
anti-tank, 14, 24, 34, 37, 41
Apollonius of Perga, 81n

Asher, R.B., 186
aspect angle, 17, 158, 220
attitude (angle), 147, 197, 232
attitude pursuit, see pursuit, attitude
augmented proportional navigation (APN), 117, 166-168, 177, 185, 187, 244-245
automobile, 16. See also motorcar
autonomous (guidance), 3
autopilot 36-37, 131, 137, 144, 172-173, 191, 197
avoidance, xv
azimuth, 27, 41, 131n

Baba, Y., 116
Baka, 49
ball, 97-8
ballistic missile, 166, 182, 187
bandwidth, 3, 38, 157, 157n, 158-159, 244
bank-to-turn (BTT), 39, 176, 229-231
barrel roll, 133n, 168
baseball, 97
bat(s), 4, 131n
beam, 4, 39
beam rider, riding, 4, 35, 39 et seq.
bearings-only, 205-206
Becker, K., 116
Bennet, R.R., 103
Bernhart, A., 48
Bernouli, Jacques, 65n
bias, 113, 166
binoculars, 41

bird, 1, 2
boat, xiii, 11-12, 39, 48, 67, 79, 144
body angle, 147, 173, 197. See also
        attitude
body control, 2-4
bomb 5, 13, 30, 48-49
Bouguer, P., 48-49, 57
Brainin, S.M., 170
Bruckstein, A., 48, 59
Bryson,A.E., 2, 184, 197, 202
bug(s), 59
bullet, 49

CADET, 160
capture zone, 54, 117, 209. See also
        launch zone
cat(s), 59n
CCD, 41
chart, 48, 69n
Chern, J.-S., 115
children, 59n
Clarinval, A., 48
classical control, 36 *et seq.*
CLOS, see command-to-line-of-sight
close range, 24. See also short range
closing speed, velocity, 6, 79, 87, 96,
        103, 159, 170, 175, 183, 205,
        218
Cochran, J.E., 115
collision, 79
collision-course (CC) conditions, 15,
        62, 78
collision-course (CC) navigation, 77
collision triangle, 78-79
command guidance, 4-6, 12, 49, 131
command-to-line-of-sight (CLOS), 4-
        5, 13-14, 39 *et seq.*, 48, 230
computer, xv, 154, 197, 209
cone, 25
constant-aspect navigation, 95-96
constant-bearing navigation, 77-78
constant projected line, 96-98

constraint, 195
control, 2, 44, 194, 196, 199, 203. See
        also guidance control
    in guidance loop, 130
    modern, see modern control
    optimal, see optimal control
control effort, 113, 117, 165, 170-171,
        184, 193, 197, 200, 211
controller, 36
control surface, 229
coplanar(ity), 7, 87, 118-119, 220
cost functional, 183
costs, xv, 104, 144
coupling 141, 229
course, 69
Criel, H.E., 170
critical range, 137-140
critical time-to-go, 138, 140
cross-product steering, 73
cross range, 31, 69
cruciform, 229
cruise missile, 231
cyclic pursuit, 59, 62

*Deckung*, 11
derivative (control), 3, 34, 37, 166,
        174
Descartes, R., 65n
deviated PP, see pure pursuit, devi-
        ated
Dhar, A., 115
differential games, xiv, 211
dog(s), 48, 59
dogfight, 48, 58, 60
Dowdle, J.R., 209
drag, 86, 149, 154, 185, 197, 231n
Dubois-Aymé, J.M.J., 48
dynamic lead guidance (DLG), 174
dynamic pressure, 131, 131n, 140, 153-
        154, 232
dynamics, 36, 43, 131 *et seq.*, 157,
        182, 192, 204

high-order, 145, 185, 191-192

first-order, 132, 152, 156, 182, 185, 187-188, 191-192, 204, 206, 208

second-order 134, 137, 182, 185, 189-192,

third-order, 191

zero-order, 3, 36, 166, 182-183, 185, 192, 198, 204

Effective navigation constant, 107 *et seq.*, 152

effort, see control effort

Eichblatt, E.J., 43

electro-optical, electro-optics, xv, 149, 156

elevation, 24, 41, 131n

endgame, 177

energy, xv

engagement plane, 47, 119, 244. See also guidance plane

Enzian, 14

equivalent noise bandwidth, 244

equivalent time-constant, 153, 205, 207

error,

    aiming, 113

    heading, 113, 133-134

    in guidance loop, 2, 31, 69

    modelling, x

    radome refraction, see radome refraction error

    scale-factor, 173, 175-176

error measuring device (EMD), 41, 172-173

escape, 47

estimator, 200, 202-203, 207-208

Euler angle, 90, 225-226

Euler's four parameters, see four parameters of Euler

evader, xiii-xiv

evading (target), 145

evolute, 94, 94n

Feed-forward, 38

Feynman, R.P., 59n

fielder, 97, 97n

field of regard, 146-147, 173

field of view (FOV), 41, 49, 145-148, 173, 197

Fi-103, 49

Fieseler, 49

filter, 130, 134, 137, 154, 173, 206-207

fin servo, 43, 131, 141

fineness ratio, 149, 154

fish, 97

fly (insect), 97

fly ball, 97, 97n

four parameters of Euler, 228

frame of coordinates (FOC), 6

    absolute, 50 *et seq.*

    inertial (IFOC), 6, 129, 194, 198, 217

    relative, 50 *et seq.*

    rotating, 219

    target's tail (TT), 17, 51 *et seq.*, 110

free fall, 22, 30

frequency domain, 37, 137

friction, 155

Fritz-X, 13, 48

frozen range, 36, 135, 140

Fueurlilie, 14

full-order, 191, 204

full-state, 189, 200

Gabriel, 5

gain, 38, 135, 152, 172-174, 188, 192, 196, 198, 232

game, see differential games

Garber, V., 185-186

Garnell, P., 43, 135, 167

Gazit, R., 143, 168

geometrical rule, 1-3

    line-of-sight, 3, 11 *et seq.*

    pure-pursuit, 47 *et seq.*

parallel navigation, 77 *et seq.*, 101, 104

German(y), 5, 11-14, 48-49, 70, 103, 231

Ghose, D., 115, 117

gimbal(s), 41, 146-147, 155-156, 172-173, 175

gravity, 113, 166

ground-to-air, xiii, 6, 13, 146, 196, 208

Guelman, M., 114, 139

guidance, xiii, 1-2
    midcourse, see midcourse guidance
    three-point, see three-point guidance
    two-point, see two-point guidance

guidance control, 182, 206

guidance law, 2-3, 129, 165
    advanced, 176, 188
    classical, xv, 165, 175, 182
    for LOS guidance, 31 *et seq.*
    for low LOS rates, 170-171 *et seq.*
    for parallel navigation, 92-94
    for pure pursuit, 69-74
    mixed, see mixed guidance law
    modern, xiv, 181-216
    related to PN, 165-180
    suboptimal, 191
    terminal, see terminal guidance

guidance loop, 2-4, 31, 129

guidance plane, 90. See also engagement plane

guidance-to-collision law, 166, 168

Gulf, 49

Gutman, S., 143, 168

gyro(scope), xv, 43, 155, 173, 176
    free, 147-148, 155, 173
    rate, 148, 207

Hawk, 4

head, see homing head

heat, 4, 49. See also IR

helix, 91

high altitude, 24, 144, 153, 233

Holt, G.C., 135

home plate, 97, 97n

homing, 6, 48, 67
    active, 4, 43
    passive, 4, 43, 205
    semiactive, 4, 43

homing head, 130

hound-hare pursuit, xiii, 47-48, 59n

Hs-293, 13, 48, 231

Hs-293D, 5, 48, 70-71

Huygens, C., 56n

Illuminate, Illumination, 4, 49

implementation, 3, 33, 121, 165, 207

infra-red, see IR

insect, xiii, 64

instantaneously planar, see planar, instantaneously

instrumentation, xv, 69, 117, 131

intercept point (IP), 54, 81, 93

interception boundary, 21

interception, locus of, 54-55, 81-82

interception zone, 20-21

involute, 94, 94n, 95

IR, 39-41, 49, 159, 205. See also heat

isochrone, 18, 52-53, 63, 80-81, 110, 238

isomaneuver, 20, 56-57

Jammer, 156

Jerger, J.J., 103

jerk, 185, 187

jink, xiv

joystick, 41

Kalman filter(ing), 154, 176, 202, 206

kamikaze, 49, 103

Kh plane, 18-20, 31-32, 235

Kim, Y.S., 123

kinematics, 6, 217
    of LOS guidance, 14, 25, 36

of parallel navigation, 78
of PP, 49
of PN, 104, 106, 113, 129

Ladies' Diary, 57
lag, 60, 112
lag-free, see dynamics, zero-order
landing, xiv, 68
Lange, E., 49
Lark, 103
laser, 4, 39, 49
latax, see lateral acceleration
launch envelope, zone, 54, 145, 209,
    210
lead angle, 15, 21, 24, 26, 37, 48, 60,
    78, 93, 166, 176, 220. See
    also kinematics
lead network, see derivative (control)
lead pursuit, 48
Leibniz, G. W., 56n
lift, 231
limaçon, 55
limit cycle, 58-59
linear-quadratic (LQ), 184, 195 *et seq.*
linear quadratic Gaussian (LQG), 200
    *et seq.*
line-of-sight guidance, xiii-xiv, 11-46,
    48, 176
    modified, 24, 29
Locke, A.S., 103
lock-on after launch (LOAL), 177
lock-on before launch (LOBL), 177
long range, 166
look angle, 130, 146-147, 149-150, 154,
    166
low altitude, 34, 144
low-pass, 130
loxodrome, 69, 69n

Mahapatra, P.R., 116-117
maneuver, see also lateral acceleration

target, xiv, 133-135, 158-159, 170,
    186, 209, 211
    exponential, 185
    evasive, 168, 187
    sinusoidal, 133, 187
maneuvering target, 57, 66, 82, 116,
    170, 185-186, 188
marine navigation, 1
marine vehicle, 3, 12, 24
mariner(s), 79
mass unbalance, 155
mate, 95
Mathews, W.E., 103
Matuszewski, J.P., 186
maximum range, 54
McGhee, R.B., 170
mechanization, 39 *et seq.*, 74, 129 *et*
    *seq.*, 165, 181, 195, 202 *et*
    *seq.*
Meir, L., 191, 204
Mercator projection, 69n
merchantman, 48, 79
microelectronics, xv
microprocessor, 182
midcourse guidance, 176-178, 181
minimum phase, 191, 191n
minimum range, 54, 145
miss distance, 108, 133-135, 145, 149,
    157-160, 166-167, 181, 184-
    192, 203-205, 211, 242
mixed guidance law, 165-166, 175-177
modern control, 33, 176
motorcar, 85, 144
Murtaugh, S.A., 170

NATO, 30
navigation, 1, 1n, 2, 69
    collision-course, see collision-course
        navigation
    inertial, 1, 176
    parallel, see parallel navigation

proportional, see proportional navigation

navigation constant 93, 101 *et seq.*, 175, 203

    effective, see effective navigation constant

near collision course (NCC), 106-107, 132, 183-187, 192, 194, 200, 244

Newell, H.E., 103

Newton, I., 56n

night insect, see insect

noise, 24, 43, 130, 135, 156-160, 172, 200, 205, 244

    angular, 156-157

    fading, 156-157

    glint, 156-159

    intensity, 156

    measurement, 201

    process, 201, 207

    receiver, 156-157

    thermal, 156

non-autonomous (guidance), 4

nonlinear(ity) 3, 43, 137, 140, 160, 196-197

nonminimum phase, 191, 191n

nonplanar, 87, 91

Nyquist criterion, 135, 138

Ogive, ogival, 149

optimal control, xiv, 38, 181 *et seq.*

optimal estimation, xiv, 182, 201-203, 206

optimal filtering, 201

optimal control guidance (OCG),182 *et seq.*

Parag, D., 143

parallelepiped, 220n

parallel navigation, xiii, 15, 30, 77-100, 138, 141, 166, 175-176

Pascal, B. and E., 55, 55n

path angle, 7, 232

Patriot, 168

Peenemünde, 103

performance index, 183-184, 193-195, 197 *et seq.*

Perrault, C., 56n

pigeon(s), 49

piracy, pirate(s), xiii, 48, 79

pitch, 129, 147, 229-230

planar(ity), 7-8, 25, 30, 47, 49, 87, 101, 104, 183, 192

    instantaneously, 7, 25, 60, 77, 87, 91

polygon, 59

Popov, V.M., 139

power spectral density (PSD), 156 *et seq.*, 206

predator, 48, 95, 103

prediction guidance law, 123

prey, xiii, 4, 48, 103, 131n

projected intercept point (PIP), 73, 94

proportional lead guidance (PLG), 165, 172-174

proportional navigation (PN), xiv, 6, 92-93, 101-180, 203, 208-209

    acceleration-compensated (ACPN), 143-144, 168

    augmented (APN), see augmented PN

    biased, 170-171

    dead-space, 171-172 *et seq.*

    generalized (GPN), 115-116

    ideal, 122

    integral form of, 173-174

    pure, see pure PN

    switched bias, 166

    synthetic, 131

    true, see true PN

    weighted, 124

proportional navigation modified by bias, 144, 166-168, 193

Prussia, xiv, 12
pure PN (PPN), 103, 109 *et seq.*
 modified, 117
pure pursuit (PP), xiii-xiv, 6, 16, 47-
 76, 80-81, 104, 165, 172
 deviated PP, 59-69, 104, 239
pursuer, xiii-xv
pursuit
 attitude, 69 *et seq.*, 172
 cyclic, see cyclic pursuit
 lead, see lead pursuit
 pure, see pure pursuit
 velocity, 69 *et seq.*, 172, 175

Quaternion, 228
Queen's University, 39

Radar, 41, 43, 49, 131, 148-149, 156,
 158-159, 205, 209
 active, 156, 205
 semi-active, 156-157
radio-frequency, see RF
radio link, 5, 14, 131
radio transmissions, 4, 49
radome, 149, 154
 ogival, see ogive
 spherical, 154n
radome refraction, 151-155
radome refraction error, 140-141, 149-
 154, 205
ramp response, 191n
range,
 close, see close range
 critical, see critical range
 cross, see cross range
 definition of, 6
 long, see long range
 short, see short range
range rate, 109
range-using TPN (RTPN), 109, 115,
 131, 143
Raytheon, 103

RCA, 103
recreational mathematics, 48, 58
re-entry vehicle 73, 197
refraction, see radome refraction
refraction coefficient, 152, 154, 205
refraction slope coefficient, 150, 205
Reitsch, H., 49
rendezvous, xiii, 79, 116, 197-200
RF (radio frequency), 43, 49
Rhinetochter, 14
rhumb line, 69
Ricatti equation, 196, 202
Riggs, T.L., 209
Roberval, G., 55n
robot(ics), xiv, 73
robust(ness), 37, 117, 195, 195n
roll, 39, 146-147, 229-230
root locus, 135
rope, 56, 58
rotational transformation, see angular
 transformation
*Rotkäppchen*, 14
Rusnak, I., 191, 204

SA-2, 30
SA-3, 30
saturation, 140, 144 *et seq.*
scalar triple product, 220
Schmetterling, 14
Schoen, E.T., 123
scopodrome, 47
screw thread, 25, 91
SD-1400-X, 13
sea skimming, 173
sea-to-sea, 5, 176
seeker, 27-28, 41 *et seq.*, 49, 71-72,
 117, 130-131, 134, 137, 145
 *et seq.*, 151, 154-156, 158-159,
 172, 175, 197
sensor, 41
separation theorem, 201
shell, 175

ship defense, 11
Short Brothers, 39
short range, 168. See also close range
Shukla, U.S., 116-117
sideslip, 231
Siemens, Werner von, xiv, 12, 12n
single-lag, see dynamics, first-order
skid-to-turn (STT), 39, 229-231
SLAM, 160
smart weapon, 49
sonar, 131n
Soviet Union, 30, 48
spacecraft, xiii, 2
Sparrow, 4
spira mirabilis, 65n
Spitz, H., 103
stability 2, 135 *et seq.*
    finite-time, 139
stabilization, 43, 141, 147, 154-156
state, 194, 200
    estimation of, 197, 200, 202-203,
      206, 209
state space, 189
steering command, 2
stochastic, 200
Stockum, L.A., 191
strapdown, 146, 165, 172-173, 175-176
structure (mechanical), xv, 144, 230
suicide pilot, 49
surface-to-air, see ground-to-air

Talos, 176
target,
    apparent, 150
    evading, see evading
    maneuvering, see maneuvering
television, see TV
terminal guidance, 177
terminal guidance (TG) law, 144, 168
three-D (3-D), 7-8, 25-26, 77, 102, 118-
    121, 133n, 185, 225
    equations of motion in, 217-224

three-point (guidance), xiii, 3-4, 6, 47,
    181
thrust, 185, 203, 230, 233
thrust vector control (TVC), 233
time domain, 33, 137
time-of-flight, 2, 53, 63, 80
time-to-go, 30, 94
    critical, see critical time-to-go
    estimation of, 205, 209
Tomahawk, 231
torpedo, xiv, 12
tracker, 24, 27, 41 *et seq.*, 145
track-via-missile, 6
tracking loop, 43, 147-148, 151, 155,
    172
tractrix, 56-57
true PN (TPN), 103, 107 *et seq.*, 132,
    167, 243
turning rate time-constant, 73, 151,
    232
turtle(s), 59
TV, 5, 41, 48-49, 70-71, 205, 230
twist-and-steer, 231
two-lag, see dynamics, second order
two-point (guidance), xiii-xiv, 4, 6, 47,
    64, 181

Underwater, 31, 43
UNESCO, 48
United States (US), xiv, 14, 41, 48-
    49, 102-103, 168, 181. See
    also America(n)

V-1, 49
variable speed, velocity, 84, 86, 140,
    141 *et seq.*, 168
velocity,
    closing, see closing speed
    radial, 7, 219
    relative, 6, 217
    tangential, 7, 219
velocity increment, 117

velocity pursuit, see pursuit, velocity
vertebrate(s), 103
Viet-Nam, 49
visual, 39

Warhead, 24
Wasserfall, 13
wavelength, 39, 41, 154
weathercock, 71
weaving (target), 133
weighting function, 72, 74, 187
Weimer, F.C., 191
Willems, G.C., 185
wind, 68, 172
wire, 14
World War II, xiv, 5, 13-14, 41, 48,
        231

X-4, 14
X-7, 14

Yang, C.-D., 115
yaw, 147, 229-230
Yuan, C.L., 103
Yuan, P.-J., 115

Zarchan, P., 160
zero-effort miss, 108, 167, 184-193
zigzag, xiv

Printed and bound by CPI Group (UK) Ltd, Croydon, CR0 4YY

03/10/2024

01040435-0011